A POCKET HISTORY
OF ARTILLERY

Light Fieldguns

A POCKET HISTORY OF ARTILLERY

Light Fieldguns

Dipl-Ing Franz Kosar

LONDON

IAN ALLAN

First Published 1971
English edition 1974

ISBN 0 7110 0485 4

© J.F. Lehmanns Verlag München 1971
© English edition Ian Allan Ltd 1974

Published by Ian Allan Ltd, Shepperton, Surrey
and printed in the United Kingdom by
The Garden City Press Limited, Letchworth, Herts

Contents

Preface

"C'est avec l'artillerie qu'on fait la guerre" — Napoleon.

The decisive role which artillery played in the wars of the 20th century is generally recognized. Nevertheless, we are still looking in vain for a comprehensive list of the fieldguns used in these actions in the relevant literature on armaments. Except for descriptions of individual guns, only the heavy and super-heavy guns have been considered.

This gap is now being closed by the **Pocket History of Artillery.** We have chosen the beginning of this century, ie the time when the gun with recoil barrel was generally adopted, as the starting point. Recoil carriage guns have been included only when they were the larger part of the equipment of the major armies after 1900, or the sole equipment of small armies. The book covers the European countries and the non-European major powers.

In this volume light fieldguns are defined as guns of up to 90mm, howitzers of up to 122mm, and all mountain guns. Further volumes deal with medium fieldguns — guns from 91 to 155mm and howitzers and mortars from 123 to 155m; heavier guns and railway guns; anti-tank guns and infantry guns; and the organization and use of artillery.

The requisite information was in some instances extremely difficult to obtain. In the field of artillery especially, the archives of various countries show regrettable gaps. In addition, a vast amount of material was lost as a result of two world wars. This applies particularly to the archives of the designers and the manufacturers. I was fortunate to receive the special support of the Austrian Study Group 'Truppendienst', in particular of Dr. F. Wiener, of the Heeresgeschichtliches Museum (Museum of Military History) and of the Kriegsarchiv (War Archives) Vienna, of the Imperial War Museum London, and of AB Bofors and Vickers Ltd. Helpful communications were received from the German Federal Archives, Koblenz, the Bibliothek für Zeitgeschichte (Library of contemporary history) Stuttgart, the Zentralbibliothek der Bundeswehr (Central library of the Federal Defence Force) Dusseldorf, the Biblioteca d'Artiglieria e Genio (Library of Artillery and Military Engineering) Rome, the Army Museum Prague, the Army Museum and War Archives Budapest, the Army Museum Bucharest, and the War Archives, Helsinki. My thanks are also due to Drs. J. Borus, W. Hummelberger, N. Krivinyi and Hw R. Surlémont for their help with obtaining technical information.

Unfortunately, it has not been possible to close all the gaps because of the sparseness of available information. I should therefore be grateful for any corrections and offers of collaboration.

Große Sperlgasse 38/16,
A — 1020 Wien

Dipl-Ing F. Kosar

Development

Introduction

Three fundamental requirements determined the whole development of fieldguns in the 20th century: increase of range, more effective ammunition, increase of mobility.

It was not only technical feasibility, but also weight limitation which for a long time stood in the way of these advances. As long as they were horse-drawn, light fieldguns were restricted in weight to about 200kg for the gun including limber, from which a maximum weight of 1400kg for the gun can be deduced. This requirement lost its importance as guns became motor-drawn. Since then the development of guns has been limited only by technological factors.

The mobility of field artillery has been considerably increased with the advent of motor traction. This called for various design modifications to the guns, especially to the carriages (sprung axles, suitable brakes, rubber-tyred wheels, etc). During the transition period before the introduction of new designs, measures of improvisation (transverse struts, trucks etc) had to be taken with the existing guns. Self-propelled guns and armoured guns represented a further stage of increased mobility. The ultimate progress in this field came with air transport, either in transport planes or by suspension from helicopters. However the latter method in particular, would again suggest the need for a reduction in the weight of guns.

Cost is naturally an important factor. Up to World War I the gun park was the most expensive item in the equipment of any army. Even later, when motorization affected ever increasing sectors of the armed forces and therefore swallowed up an increasing proportion of the available funds, guns continued to form an important item. They have, however, a long life. The French fieldgun, 75mm, M 97 (Schneider) for instance, was the mainstay of the French army and thus of a Great Power for nearly 50 years. Perhaps this was a unique case.

The question of cost is also the driving force behind the frequent demand for new designs able to utilise existing stocks of ammunition, ie retention of the same calibre. Because of the huge stocks of ammunition this has a far greater bearing on fieldguns than on the heavy guns. Efforts to introduce standard carriages for fieldguns and field howitzers should also be mentioned; they were part of the trend, which began during World War I and which continued during the inter-war years towards a standardization of artillery equipment. The logical next step was the standardization of ammunition achieved during World War II and afterwards for field, tank and anti-aircraft guns.

The two world wars are naturally the most important periods of the entire era. On the one hand they acted as powerful stimulants to development, but on the other they left behind, as a result of the greatly increased war production, huge stocks which had to be retained for future use. In the field of mass production especially, great national differences in conception became evident. Whereas Germany, during both wars, endeavoured to increase artillery performance by introducing numerous new designs, France during World War I and the US during World War II continued to mass produce large numbers of existing models unchanged until the end of hostilities. The advantage in sheer numbers of available guns was obvious.

Light fieldguns

Light fieldguns have mainly had calibres between 75 and 77mm since the end of the 19th century. Great Britain was the sole exception with 83.8mm. During World War I the Central Powers, too, experimented with larger calibres, but did not progress beyond prototypes. The difficulties of ammunition supply prevented their general introduction.

Attempts to increase the range were based on various principles, but increased muzzle velocity (V_o) was the most widely used. At the turn of the century the average was 500m/sec. Russia was the exception with values close to 600m/sec. Between the two world wars velocities rose to about 700m/sec. To meet the demands of anti-tank defence a further increase was achieved during World War II, with top values of 800m/sec. By then, the light fieldgun had already become an anti-tank gun.

A second possibility of increasing the range lay in the increase of the maximum angle of elevation. Up to World War I, elevation extended generally from $-10°$ to $+20°$ increasing to about $40°$ during that war. This value has not been exceeded since. Some of the older guns were modified accordingly. The only exceptions were fieldguns designed during the inter-war period for angles of elevation of up to $80°$, for additional use as anti-aircraft guns.

The third possibility, improved shape of the projectile, is not as significant with light fieldguns as with heavy guns. Before World War I the maximum range of the conventional light field gun of 75—77mm calibre was between 5 and 7km. By 1918 it had increased to 11km, and by the outbreak of World War II to 14km. The larger-calibre fieldguns naturally had greater range

The effectiveness of the projectile was increased mainly by the introduction of larger calibres. It was this measure which led to the substitution of the light fieldgun by field howitzers or gun howitzers, since a larger gun calibre was ruled out for reasons of weight limitation. A second way was to use longer shells or shells with thinner casings to accommodate a larger quantity of explosive. But both possibilities were restricted by ballistic conditions and could be realized only within certain limits.

The increase in the traversing range, too, affected the use of the light fieldgun considerably. As the direct descendants of the recoil carriage guns, the first recoil barrel guns had a traversing range of only 4° to 8°. But the first fieldgun with split-trail carriage, the fieldgun, 75mm M 11, (Deport), with a traversing range of 54° had been introduced even before World War I. It showed that the box trail carriage offered no appreciable advantage in weight over the split-trail version. The radically new idea of this gun was demonstrated by its subsequent development: a traversing range of 60° was never exceeded.

The light fieldgun had already passed the peak of its development during World War I. Whereas before 1914 about 90 per cent of the light batteries throughout the world were still equipped with light fieldguns, later this proportion progressively decreased. At the beginning of World War II, only in France were considerable sections of the artillery still equipped with light fieldguns. All other countries had switched increasingly to the light field howitzer. Since the end of World War II light fieldguns survive only in the Eastern Bloc, where they are however, used mainly as anti-tank weapons.

Gun howitzers

The gun howitzer is the result of efforts to combine the ballistic performance of the gun with the elevation range of the howitzer in order to provide both flat-trajectory and high-angle fire. They can therefore be regarded as a compromise between these two types of gun. The contents of the two relevant sections apply generally.

Although the arms industry had already developed several prototypes during the years following World War I the first gun of this kind was not introduced until the end of the twenties. It was the Schneider gun howitzer, 85mm M 27, and was supplied to the Greek army. The next country to follow was Great Britain, whose gun howitzer, 87.6mm Mk I (Royal Ordnance/Vickers) clearly shows its descent as its barrel was a re-bored fieldgun barrel. The last country to adopt gun howitzers was the Soviet Union. Although in the case of the medium calibres this step had already been taken before World War II, light gun howitzers were not introduced until the sixties; but they appeared in the form of perhaps the most effective current light fieldgun, the gun howitzer, 122mm, M 63 (state) with universal carriage.

Light field howitzers

With regard to the calibre of the light field howitzer, two groups had

9

formed before World War I. Germany and Austria-Hungary chose a calibre around 100mm, whereas France, Great Britain and Russia adopted about 120mm. The smaller countries decided less on technical aspects than on their political ties with one of the two groups. In the end, the most widely accepted calibre was 105mm. Surviving stocks from the Austro-Hungarian army were responsible for the use of the 100mm calibre far into World War II; Great Britain adopted a completely different solution; the Soviet Union was the only power to retain the 122mm calibre.

An increase in the range of the light field howitzer was possible only through an increase in the muzzle velocity; it was similar to that of the light fieldgun and often led to the development of gun howitzers. The light field howitzers around 100mm had a range of about 6km before World War I, 10km towards the end of that war, and 12 to 13km during World War II. Today the maximum range has reached 15km. The performance of the light field howitzers around 120mm was of course slightly superior.

In the increase of the traversing range, too, the light field howitzers followed the light fieldguns. Since the end of World War II universal gun carriages have come into use. Their development began in USA after 1918 and was resumed by Germany during World War II. But in both countries no progress was made beyond the design stage and wooden mock-ups. The first universal gun carriages were built in France for the field howitzer 105mm M 50 (DTAT). They were subsequently introduced by Sweden (field howitzer, 105mm, 4140 Bofors) and the Soviet Union (gun howitzer 122mm, M 63, state). Universal gun carriages, which weigh almost twice as much as the split-and box-trail carriages, could of course be adopted only after the advent of motorization.

During the inter-war period the light field howitzer progressively took the place of the light fieldgun, but it was unable to meet the requirements of World War II because it was unsuitable as an anti-tank weapon. In the West today it is found only occasionally at divisional level, mainly in the form of US stock from World War II or, as in Sweden, on a universal carriage. In the vast majority of cases howitzers of larger calibre (155mm) have taken its place. The East, with its modern gun howitzer, has adopted its own solution.

Mountain Guns

The development and introduction of mountain guns depends mainly on geographical conditions. This explains why the most successful models were developed by Austria-Hungary and later by it successor state Czechoslovakia. Guns built during World War I were still used by many armies during World War II. Further development after World War I was parallel to that of the light fieldguns.

The special problem here is the required number of loads for transportation. In the beginning the weight of the gun and hence the number of loads, was kept low, even at the expense of fire power. The increase in performance subsequently demanded, led to an increase not only in the calibre, but also in the number of loads, from three to four before World War I (65mm to 76.2mm) to twelve today (105mm).

The number of mountain guns has steadily decreased in all armies. Both large-calibre mortars and recoilless guns have contributed to their displacement. They have become largely redundant, chiefly because of the building of new roads in the mountainous regions. In addition, even light field howitzers can now be transported to any site in the mountains by helicopters. Today only one mountain gun, the Italian mountain howitzer, 105mm M 56 (OTO Melara) is widely used by various armies. It is also employed by airborne units, both for air drop and helicopter transport.

Equipment of the Individual Countries

Albania

Without an industry of her own Albania has never been able to equip her army with guns of her own manufacture.

After the country became independent in 1913 the armed forces obtained only obsolete equipment. During the inter-war years it was supplied by various sources and therefore not standardized. After World War II the Albanian army was completely equipped with Soviet equipment, but since relations between Albania and Russia soon deteriorated, the Albanian army today has only guns from World War II:

fieldgun, 76.2mm, M 42 (state/Soviet Union)
field howitzer, 122mm, M 38 (state/Soviet Union)

Austria-Hungary/ Austria

The majority of the guns of the Austro-Hungarian army were made by Škoda of Pilsen. In addition, small numbers of mountain guns made by Artillerie-Zeugs-Fabrik, Vienna were introduced. The firm Böhler, Kapfenberg developed a field howitzer that was superior to the contemporary Škoda model even before World War I, but it was not introduced because of production difficulties. Böhler fieldguns reached units in the field only during World War I. Of the older guns, many were built as 'war department models'.

At the beginning of World War I Austria-Hungary was the only Great Power whose field artillery was still largely equipped with recoil carriage guns. These were the field howitzer, 104mm, M 99 (state,Austria-Hungary) and the mountain gun, 72.5mm, M 99 (state, Austria-Hungary).

In addition, however, modern guns, mainly light fieldguns and a small number of mountain guns and mountain howitzers, had also been introduced:

fieldgun, 76.5mm, M 05 (state/Škoda)
fieldgun, 76.5mm, M 05/08 (state/Škoda)
mountain gun, 72.5mm, M 08 (Art.-Zeugs-Fabrik)
mountain gun, 72.5mm, M 09 (Škoda)
mountain howitzer, 104mm, M 09 (Škoda)
mountain howitzer, 104mm, M 10 (Škoda)

New guns, intended as replacements for the obsolete recoil carriage guns, were already being developed, but had not yet been introduced in 1914. At the beginning of the war, guns which had been ordered by other countries, were retained and used by the Austro-Hungarian army:

fieldgun, 75mm, M 11 (Škoda)
field howitzer, 105mm, M 12 (Škoda)
field howitzer, 104mm, M 14 (Škoda)
mountain gun, 75mm, M 14 (Škoda)

During World War I, from the beginning of 1915, the new guns (already mentioned) were introduced.

fieldgun, 76.5mm M 17 (Škoda)
fieldgun, 76.5mm, M 18 (Böhler)
field howitzer, 100mm, M 14 (Škoda)
mountain gun, 75mm, M 15 (Škoda)
mountain howitzer, 100mm, M 16 (Škoda)

By the end of the war the prototype was completed of the heaviest mountain gun, to reinforce the mountain artillery—the mountain howitzer 150mm M 18 (Škoda). In addition there were several German mountain guns—mountain gun 75mm M 14 (Rheinmetall).

Of all these guns the Austrian Republic continued to use only the following:

fieldgun, 76.5mm, M 17 (Škoda)
fieldgun, 76.5mm, M 18 (Böhler)
field howitzer, 100mm, M 14 (Škoda)
mountain gun, 75mm, M 15 (Škoda)
mountain howitzer, 100mm, M 16 (Škoda)

Indeed, no new field or mountain guns were introduced up until the beginning of World War II, but by the use of a new type of ammunition—the projectile M 32—the range of the fieldgun was increased substantially.

After World War II, on the formation of the Army of the Second Republic, a completely new armament was necessary since no older guns were left. Equipment left by the former Occupying Powers had therefore to be used. Later a few German guns of World War II were bought abroad, but are now used only by reserve units:

fieldgun, 76.2mm, M 42 (state/Soviet Union)
field howitzer, 105mm, M 2A1 (Ordnance Department)
field howitzer, 105mm, M 18/40 (Rheinmetall)

The Soviet gun has now been scrapped.

Belgium

In the firms of Cockerill and FRC, Belgium had two manufacturing plants for guns, but nevertheless decided to import guns from other countries. Only during and after World War I did these two manufacturers produce their own designs, based on modifications. Up till then they had built only the foreign-made fieldgun introduced in the Belgium army under licence.

Before World War I the Belgian field artillery was equipped with the fieldgun, 75mm, M 05 (Krupp).

Immediately before the outbreak of that war the field howitzer, 120mm (St Chamond) was introduced. A large part of this equipment was lost at the beginning of World War I. For re-equipment, the French fieldgun, 75mm, M 97 (Schneider) and the field howitzer 120mm, M 11 (Schneider) in addition to captured German guns was adopted.

After the end of the war the Belgian army was re-equipped. The remaining Krupp guns were modernized, fieldgun 75mm, TR (FRC/Krupp) and the captured or surrendered German fieldguns, 77mm, M 16 (Rheinmetall) rebuilt as:

fieldgun, 75mm, GPI (Cockerill/Rheinmetall)
fieldgun, 75mm, GPII (Cockerill/Rheinmetall)
fieldgun, 75mm, GPIII (Cockerill/Rheinmetall)

To the remaining stock of Belgian-built field howitzers, captured German field howitzers, 105mm, M 16 (Rheinmetall) were added. With those four guns the Belgian army entered World War II.

After World War II equipment was standardized as in most other NATO countries with the adoption of the US field howitzer, 105mm, M-2A1 (Ordnance Department).

In recent years there has also appeared the mountain howitzer, 105mm, M 56 (OTO Melara).

Bulgaria

Like all other Balkan countries, Bulgaria, too, was unable to design and develop her own guns owing to the absence of heavy industry. Before World War I, the entire gun park was bought in Germany and France. After World War II the Bulgarian People's Army was equipped with guns by the Soviet Union.

Before World War I the Bulgarian army relied on French fieldguns and German and French mountain guns:

fieldgun, 75mm, M 05 (Schneider)
field howitzer, 120mm, M 07 (Schneider)
field howitzer, 120mm, M 11 (Schneider)
mountain gun, 75mm, M 05 (Krupp)
mountain gun, 75mm, M 07 (Schneider)

Although the models were the same, guns bought later to supplement the existing equipment had the year of delivery stamped on them. Thus the French fieldgun is also called M 07 and M 08, the German mountain gun M 08 and M 10.

During World War I Germany and Austria-Hungary supplied Bulgaria with guns for her newly recruited troops. They consisted of:

fieldgun, 77mm, M 96 new version (Krupp/Ehrhardt)
fieldgun, 77mm, M 16 (Rheinmetall)
field howitzer, 105mm, M 98/09 (Krupp)
field howitzer, 105mm, M 12/16 (Krupp)
field howitzer, 105mm, M 16 (Rheinmetall)
mountain gun, 75mm, M 15 (Škoda)
mountain howitzer, 105mm (Krupp)

Practically all of these guns remained in use up to the beginning of World War II. Only one new mountain gun, the mountain gun 75mm M 36 (Bofors) was introduced.

After World War II the Bulgarian army received Soviet fieldguns:

fieldgun, 76.2mm, M 42 (state, Soviet Union)
fieldgun, 85mm, M 45 (state, Soviet Union)
field howitzer, 122mm, M 38 (state, Soviet Union)

Modern weapons such as the gun howitzer, 122mm, M 63 (state, Soviet Union), have not yet been seen in Bulgaria.

Czechoslovakia

After the break-up of Austria-Hungary, Czechoslovakia was in the fortunate position of having the arsenal of the old monarchy, the firm of Škoda Pilsen on her territory. Here, the experiences of World War I were evaluated, and, from the very beginning, development work was carried out for the Czechoslovak army and for export. The company was nationalized after World War II.

The newly established Czechoslovak army was initially equipped with guns from the stocks of the Austro-Hungarian forces. They had all been manufactured by Skoda.

fieldgun, 75mm, M 11 (Škoda)
fieldgun, 76.5mm, M 17 (Škoda)
field howitzer, 100mm, M 14 (Škoda)
mountain gun, 75mm, M 15 (Škoda)
mountain howitzer, 100mm, M 16 (Škoda)

But as early as the beginning of the twenties improvements of the existing models were introduced. They were successful mainly in the howitzers, in which modifications increased performance:

field howitzer, 100mm, M 14/19 (Škoda)
mountain howitzer, 100mm, M 16/19 (Škoda)

This light field howitzer was the backbone of the Czechoslovak divisional artillery right up to 1938. The improvement of the fieldgun was less successful, so that the current model continued in production, in small numbers, up to 1938.

The next wave of modernization did not begin until the thirties. As a result of the experiences with the models designed for export, new fieldguns were introduced in the Czechoslovak army. Although development was complete as early as 1930, the guns reached the units only from 1935 onwards. Here, too, a standard carriage was decided upon:

fieldgun, 76.5mm, M 30 (Škoda)
field howitzer, 100mm, M 30 (Škoda)

The demand for the retention of the existing stocks of ammunition once again stood in the way of optimum solution. In addition to these two guns, a new gun was designed for the Horse Artillery; its introduction, however, had to be postponed in favour of that of other, more urgent, equipment; the fieldgun, 75mm, M 35 (Škoda).

The last models of Czechoslovak guns underwent trials in 1938, and preparations for their mass production were complete. But owing to the occupation of the country by the Germans these projects never came to fruition. The light field howitzer was intended to become the main weapon of the divisional artillery:

fieldgun, 76.5mm, M 39 (Škoda)
field howitzer, 100mm, M 39 (Škoda)

After World War II abandoned German guns, especially the field howitzer, 105mm, M 18/40 (Rheinmetall), were used as a stopgap.

After the establishment of the People's Democracy, the switchover to Soviet guns was rapid:

fieldgun, 76.2mm, M 42 (state/Soviet Union)
fieldgun, 85mm, M 45 (state/Soviet Union)
field howitzer, 122mm, M 38 (state/Soviet Union)

After the initial integration of the Czechoslovak industry into the production of Soviet guns, a Czech-designed gun was soon again adopted by the Czech army—the fieldgun, 85mm, M 52 (state/ČSSR). This gun supplements or replaces the Soviet model of the same calibre.

Denmark

To date Denmark has not designed any guns of her own, and her army has to rely on foreign purchases for its equipment.

Up to the beginning of World War II the only light fieldgun of the Danish forces was the fieldgun, 75mm, M 02 (Krupp) introduced at the beginning of the century.

After World War II German fieldguns, abandoned in the country, were at first used. In the course of standardization, however, they were subsequently scrapped in favour of the British gun howitzer, 87.6mm, Mk II (Royal Ordnance) and later the US field howitzer, 105mm, M-2A1 (Ordnance Department).

Estonia

Without any national arms industry, the Estonian army was equipped by France and Great Britain after the country had gained its independence. The gun park accordingly comprised:

fieldgun, 75mm, M 97 (Schneider)
fieldgun, 76.2mm, M 04 (Vickers)
fieldgun, 83.8mm, Mk IV (Vickers)

Finland

Finland did not have an arms industry of her own until World War II. Since then Tampella have manufactured fieldguns by fitting new barrels on captured Soviet carriages.

During the War of Liberation 1917/18 the newly created Finnish army was equipped mainly with captured Russian guns:

fieldgun, 76.2mm, M 02 (Putilow)
field howitzer, 122mm, M 10 (Schneider)
mountain gun, 76.2mm, M 09 (Schneider)

In addition, Germany supplied the fieldgun, 77mm, M 16 (Rheinmetall).

At the beginning of the twenties new guns were bought in France, ie the fieldgun, 75mm, M 22 (Schneider).

Only during the Winter War 1939/40 and its continuation 1941/44 did modern guns reach the army, mainly in the form of captured Soviet guns:

fieldgun, 76.2mm, M 36 (state, Soviet Union(
field howitzer, 122mm, M 38 (state, Soviet Union)

At this time the Finnish arms industry made its first mark, beginning with the rebuilding of captured guns. Above all, existing barrels were mounted on the carriage of the fieldgun, 76.2mm, M 36 (state, Soviet Union). This produced the field howitzer, 105mm, M 37-10 (Tampella).

In addition, Germany supplied many guns to the Finnish army, among others the field howitzer, 105mm, M 18 (Rheinmetall) under the designation 105mm field howitzer M 33.

After World War II the carriages of the captured Soviet fieldguns, 76.2mm, M 36 (state, Soviet Union) continued in use, but with a barrel newly developed by Tampella. This gun is called field howitzer, 105mm, M 61-37 (Tampella).

France

Like Germany, France, had, since the end of the last century, an arms factory of international importance, which, naturally, also equipped the French army. Together with the German firm of Krupp, Schneider dominated the world's arms markets. Between the two world wars, this company continued to supply large quantities of guns to foreign armies. The firm of St Chamond, on the other hand, produced only a small number of fieldguns for the French army and for export. Various state arsenals also contributed to the production. After World War II, DTAT assumed responsibility for development. Various designers, such as de Bange, Baquet, and Deport made a name for themselves, the last-named becoming known mainly through his split trail carriage.

At the end of last century France was the first country to adopt a recoil barrel gun for its field artillery. The quality of this design can best be judged by the fact that this gun was introduced by many countries and remained in use for almost half a century. As late as the thirties the US army adopted the barrel of this gun on a new carriage. In addition to this fieldgun France used a 'short gun' of 19th century vintage up to World War I. The main equipment in 1914 therefore consisted of the fieldgun, 75mm, M 97 (Schneider) and the field howitzer, 120mm, M 90 (Baquet).

There was, in addition, a small number of two other fieldguns in use:

fieldgun, 75mm, M 12 (Schneider)
fieldgun, 75mm, M 14 (Schneider)

At the outbreak of the war, the existing gun park was augmented by the re-introduction of a large number of obsolete guns—the fieldgun, 90mm, M 77 (de Bange).

The three last-named models were, however, soon scrapped.

During World War I, only two new guns reached units in the field—the fieldgun, 75mm, (St Chamond) and the field howitzer, 120mm, M 15 (Schneider), but were soon replaced by the fieldgun M 97. This was the standard weapon of the French field artillery right up to World War II.

The mountain gun used during World War I was the mountain gun, 65mm, M 06 (Schneider-Ducrest). Owing to 'ack of demand it was hardly used.

After World War I only the equipment of the mountain artillery was replaced, whereas the fieldgun with the improved ammunition, introduced during the war, continued in use. In 1939 the mountain artillery was equipped with:

mountain gun, 75mm, M 19 (Schneider)
mountain gun, 75mm, M 28 (Schneider)
mountain howitzer, 105mm, M 19 (Schneider)
mountain howitzer, M 28 (Schneider)

The rebuilding of the fieldgun with split trail carriage, the fieldgun 75mm, M 97/33 (Schneider) was not successful and was sold to Brazil. Before World War II new field howitzers were also introduced:

field howitzer, 105mm, M 34 (Schneider)
field howitzer, 105mm, M 35 (Atelier Bourges)

After World War II the artillery had to be completely re-equipped as the existing gun park had been lost in 1940. Here, too, US material, the field howitzer, 105mm, M-2A1 (Ordnance Department) was adopted.

Very soon, however, the French began to design their own guns again, but financial reasons prevented the introduction of the new field howitzer, 105mm M 50 (DTAT). This gun, too, might have played a pioneering role in the branch of field artillery. It was, after all, the first field howitzer on a universal carriage.

The mountain howitzer, 105mm, M 56 (OTO Melara) has recently been introduced into the French Forces.

Germany

Deutsches Reich/Federal Republic

In the firm of Krupp Germany had, during the first half of the century, one of the most important arms manufacturers in the world. In continuous, fierce competition with the French firm of Schneider, Krupp was one of the main suppliers of guns throughout the world. Before World War I Krupp was joined by the firm of Ehrhardt, later known as Rheinmetall.

At the beginning of the century Germany was forced to replace her recently introduced guns—the fieldgun, 77mm, M 96 (Krupp) and the field howitzer, 105mm, M 98 (Krupp)—because as recoil carriage models they could no longer compete with the recoil barrel guns already generally adopted abroad. By converting them into recoil barrel guns the Germans were, however, largely able to make good this deficiency by about 1910 with the introduction of the fieldgun, 77mm, M 96 new type (Krupp/Ehrhardt), the field howitzer, 105mm M 98/09 (Krupp) and the mountain howitzer, 105mm, M 40 (Böhler).

These were the two guns with which the German field artillery entered World War I. New guns were not introduced until the middle of that war:

fieldgun 77mm, M 16 (Rheinmetall)
field howitzer, 105mm, M 16 (Rheinmetall)

Since the new fieldgun was too heavy for certain purposes, production of the fieldgun, 77mm, M 96 new version (Krupp/Ehrhardt) was resumed during the war under the designation fieldgun, 77mm, M 96/16 (Krupp).

In addition, numerous experimental guns were developed and tried, several Škoda models among them. But only the field howitzer, 105mm, M 17 (Krupp) was actually introduced.

Before the beginning of World War I the Germans had no proper mountain artillery. The mountain gun, 75mm, M 08 (Rheinmetall) was produced in small numbers, but only for use in the colonies.

When, after the beginning of the war, the need for mountain guns arose export models were at first retained for use by the German Army:

mountain gun, 75mm, M 13 (Krupp)
mountain gun, 75mm, M 14 (Rheinmetall)

During the war two new developments were added to these guns:

mountain gun, 77mm, M 15 (Rheinmetall)
mountain howitzer, 105mm (Krupp)

In addition, the French infantry gun, 37mm, M 16 (Hotchkiss) was rebuilt to become the mountain gun, 37mm (Krupp/Hotchkiss).

Later all these mountain guns were largely replaced by the mountain gun, 75mm, M 15 (Škoda) in the course of the standardization of the gun park of the Central Powers.

After the end of World War I only two models were retained as standard artillery equipment of the Reichswehr:

fieldgun, 77mm, M 16 (Rheinmetall)
field howitzer, 105mm, M 16 (Rheinmetall)

The light field howitzer remained in use with some reserve units until the end of World War II.

Not until 1935 were new guns introduced in large numbers. They had already been developed in the twenties, but were mostly designated M 18 as disguise (Treaty of Versailles):

fieldgun, 75mm, M 16 new type (Krupp)
fieldgun, 75mm, M 18 (Krupp)
fieldhowitzer, 105mm, M 18 (Rheinmetall)

All three guns were used until 1945; the light field howitzer was the standard weapon of the divisional artillery.

After the beginning of World War II some new guns were introduced, partly designed for other countries, and partly using components of such guns:

fieldgun, 75mm, M 38 (Krupp)
field howitzer, 105mm, M 18/39 (Rheinmetall/Krupp)

The field howitzer, 105mm, M 18M (Rheinmetall/Krupp) must be mentioned as an improvement of the existing field howitzer.

To meet the demands by units in the field for increased effectiveness of the divisional artillery against enemy armour, modified guns were introduced in increasing numbers during the second half of the war. Thus barrels of the light field howitzer were mounted on anti-tank carriages, and anti-tank barrels on field howitzer carriages. This produced the fieldgun, 75mm, 7 M 85 (Rheinmetall) and the field howitzer, 105mm, M 18/40 (Rheinmetall).

Other designs, such as the field howitzers, 105mm M 43 (Krupp and Škoda) on a universal carriage were built as prototypes only or did not progress beyond the mock-up stage.

New mountain guns again were late in reaching the field units. In addition to two original designs, mountain guns of the Austrian Federal Army and Czechoslovakian army stock were taken over:

mountain gun, 75mm, M 15 (Škoda)
mountain gun, 75mm, M 36 (Rheinmetall)
mountain howitzer, 100mm, M 16 (Škoda)
mountain howitzer, 105mm, M 40 (Böhler).

After World War II, no old guns were available when the German army was re-established. Thus, the divisional artillery was equipped mainly with the US field howitzer, 105mm, M-2A1 (Ordnance Department).

This gun, however, was soon improved by mounting a longer barrel on the American carriage and by the introduction of several further new features, and was designated field howitzer, 105mm, L (Rheinmetall).

In addition, the Italian mountain howitzer, 105mm, M 56 (OTO Meiara) was bought for the equipment of airborne and mountain units.

At present a new 105mm gun is being developed in co-operation with British firms.

German Democratic Republic

When the National People's Army was established it was equipped with Soviet material:

fieldgun, 76.2mm, M 42 (state, Soviet Union)
fieldgun, 85mm, M 45 (state, Soviet Union)
field howitzer, 122mm, M 38 (state, Soviet Union)

The 76.2mm fieldgun has now been scrapped, and replaced by the fieldgun, 85mm, M 45 (D-48) (state, Soviet Union) with auxiliary drive and the new gun howitzer, 122mm, M 63 (state, Soviet Union) on universal carriage.

With this equipment the main force of the artillery, like its Soviet example, is suitable for anti-tank defence.

Great Britain

Like most Great Powers, Great Britain had a famous arms factory by the name of Vickers. This company, however, acquired its international importance through the construction of naval rather than fieldguns. Government arsenals and the firm of Armstrong co-operated with Vickers and contributed to the production. Before World War II the development of new guns was transferred to the Armaments and Development Establishment.

At the beginning of the century the British army was equipped with obsolete recoil carriage guns. A German fieldgun, the fieldgun, 76.2mm, M 00 (Ehrhardt) was therefore bought for the Boer War.

Soon, however, British designs were introduced; theirs was the largest calibre chosen in those days by any country for the artillery of the division. The horse artillery, on the other hand, was equipped with guns of conventional calibres:

fieldgun, 76.2mm, M 04 (Vickers)
fieldgun, 83.8mm, M 03 (Vickers)

In addition a light field howitzer, already introduced at the end of last century, was used, but replaced by a more modern weapon before the beginning of World War I:

field howitzer, 127mm, M 97 (Great Britain)
field howitzer, 114mm, M 10 (Vickers)

The mountain artillery was employed mainly in colonial wars (Sudan, Tibet) and used mountain gun, 70mm, M 11 (Vickers) and mountain howitzer, 75mm, M 01 (Vickers).

These were the guns with which Great Britain entered World War I.

During World War I only minor improvements of existing types were carried out. New guns were introduced only towards and after the end of the war:

fieldgun, 83.8mm, Mk IV (Vickers)

field howitzer, 114mm, M 17 (Vickers)
mountain howitzer, 93.9mm, M 18 (Vickers)

Later, an improved fieldgun with split-trail carriage, the fieldgun, 83.8mm, Mk V (Vickers) was added.

During the inter-war period the entire field artillery was motorized, and the two latest models of the fieldgun (Mk IV and Mk V) and the field howitzer were modified for this purpose.

Immediately before the beginning of World War II the British introduced a new gun, which incorporated rebored barrels of the existing fieldgun. The result was the gun howitzer, 87.6mm, Mk I (Royal Ordance/Vickers).

Soon after the outbreak of World War II a newly developed gun howitzer was added, the gun howitzer, 87.6mm, Mk II (Royal Ordnance/Great Britain).

Until long after the end of that war it remained the standard equipment of the light field artillery.

During World War II, a US gun, the mountain howitzer, 75mm, M-1A1 (Ordnance Department) was adopted for airborne units, and scrapped only in 1960.

About that time the airborne artillery and part of the field artillery began to receive the Italian mountain howitzer, 105mm, M 56 (OTO Melara).

In 1974 the gun howitzers in use at present are to be replaced by the new fieldgun, 105mm (state).

Greece

Since Greece has no arms industry of her own, she depends on the purchase of guns from abroad. Up to the beginning of World War II, she obtained all her guns from France as a result of her political ties. Only the Greek Colonel Danglis made a name for himself as a designer. The mountain gun he developed was built by Schneider in France.

Before World War I the following light fieldguns and mountain guns were bought:

fieldgun, 75mm, M 06 (Schneider)
mountain gun, 75mm, M 06/09 (Schneider/Danglis)

After World War I the existing gun park was supplemented by other French guns:

fieldgun, 75mm, M 97 (Schneider)
gun howitzer, 85mm, M 27 (Schneider)
mountain gun, 75mm, M 19 (Schneider)

All the Greek guns were lost in the course of World War II. After 1945 various arms captured or supplied by the Allies were used, of which only the US field howitzer, 105mm, M-2A1 (Ordnance Department) was retained as standard equipment.

Hungary

Fieldguns were already being built in Hungary during World War I. They were, however, not for original design, but manfuactured to plans by Škoda in the works of the Hungarian State Railways, MÁVAG, where later guns were built for the Hungarian army; but the main activity was the modernization of obsolete models.

Initially the equipment of the newly established Hungarian army in 1918/19 naturally drew on stocks of the Austro-Hungarian army and included:

fieldgun, 76.5mm, M 05/08 (state/Škoda)
fieldgun, 76.5mm, M 18 (Böhler)
field howitzer, 100mm, M 14 (Škoda)
mountain gun, 75mm, M 15 (Škoda)

The gun park was not modernized until the thirties. The first to be rebuilt were the mountain guns. This produced the following models:

mountain gun, 75mm, M 15/31 (MÁVAG/Škoda)
mountain gun, 75mm, M 15/35 (MÁVAG/Škoda)

In addition, the field howitzer, 105mm, M 18 (Rheinmetall) was bought in Germany and introduced by the army as M 37.

During World War II large numbers of this field howitzer, as well as a few of the Hungarian-designed field howitzer, 105mm, M 40 (MÁVAG, were added.

After World War II the remaining old guns were replaced by the following Soviet models, some of which are still in use today:

fieldgun, 76.2mm, M 42 (state, Soviet Union)
fieldgun, 85mm, M 45 (state, Soviet Union)
field howitzer, 122mm, M 38 (state, Soviet Union)

The two light fieldguns enable the Hungarian army effectively to fight enemy armour according to Soviet practice. More recent guns such as the gun howitzer, 122mm, M 63 (state, Soviet Union) have not yet been supplied to the Hungarian People's Army.

Irish Republic

In the absence of any heavy industry the Irish army was equipped with British material. After the establishment of the Irish Free State the army received the following guns:

fieldgun, 83.8mm, Mk IV (Vickers)
field howitzer, 114mm, M 10 (Vickers)
mountain howitzer, 93.9mm, M 18 (Vickers)

During World War II these obsolete guns were replaced by the gun howitzer, 87.6mm, Mk I (Royal Ordnance/Vickers) no longer in service with the British army. The Irish artillery units are still equipped with this gun. Its replacement with a new 105mm field howitzer is planned.

Italy

For a long time Italian industry was in no position to supply the army with all the guns it required. OTO and Vickers-Terni exclusively produced foreign designs under licence. Turin Arsenal developed only mountain guns. Not until the middle thirties did Ansaldo produce their own fieldgun models. After World War II this company was replaced by OTO Melara.

Before World War I the Italian army depended almost exclusively on foreign guns. Another disadvantage was the decision by the Italian army at the beginning of the century to introduce a recoil carriage gun, the fieldgun, 75mm, A. Since at that time even minor powers already used recoil barrel guns, production of this gun was soon halted and foreign guns were bought or built under licence:

fieldgun, 75mm, M 06 (Krupp)
fieldgun, 75mm, M 11 (Krupp)
fieldgun, 75mm, M 11 (Deport)

Only the mountain artillery was equipped with Italian-designed guns, an obsolescent recoil carriage gun and a recoil barrel gun introduced shortly before the beginning of the war:

mountain gun, 70mm, M 02 (Turin Arsenal)
mountain gun, 65mm, M 13 (Turin Arsenal)

During the war French guns, such as the mountain gun, 70mm, M 08 (Schneider) were also introduced.

After World War I the army was equipped largely with captured or surrendered guns, (some of which were later rebuilt), of the Austro-Hungarian army:

field howitzer, 100mm, M 14 (Škoda)
mountain gun, 75mm, M 15 (Škoda)
mountain howitzer, 100mm, M 16 (Škoda)

In addition the guns already in use before World War I were retained except the obsolete mountain gun, 70mm, M 02.

Only during the second half of the thirties were new guns of Italian design introduced:

fieldgun, 75mm, M 37 (Ansaldo)
field howitzer, 75mm, M 35 (Ansaldo)
mountain gun, 75mm, M 34 (Ansaldo)

Their numbers were, however, insufficient for a complete re-equipment of the units, so that during World War II the older guns were also used.

After World War II the Italian army was equipped with the British gun howitzer 87.6mm, Mk II (Royal Ordnance) and later the US field howitzer, 105mm, M-2A1 (Ordnance Department). Later, the Italian-designed mountain howitzer, 105mm, M 56 (OTO Melara) was added for use by infantry, mountain and airborne units.

Japan

Before World War I, the Japanese industry was not capable of developing modern guns for the Japanese army. But it soon began to build or rebuild foreign guns under or without licence. It was mainly the government Osaka Arsenal which carried out this work according to the plans of General Arisaka. After World War I all the guns were produced in Japan, partly according to plans of Schneider, France.

Before World War I German fieldguns, or German guns built under licence in Japan or copied with Japanese improvements were exclusively used by the army:

fieldgun, 75mm, M 05 (Krupp)
fieldgun, 75mm, M 05 (Arisaka, Osaka Arsenal)
fieldgun, 75mm, M 08 (Osaka Arsenal/Krupp)
field howitzer, 120mm, M 05 (Krupp)
mountain gun, 75mm, M 08 (Krupp)

After World War I a fieldgun and the mountain gun were rebuilt or newly built in improved versions:

fieldgun, 75mm, M 05 (improved) (Osaka Arsenal)
mountain gun, 75mm, M 08 mod. (Osaka Arsenal)

Only during the early thirties was the gun park further modernized. It was now Schneider that provided licences for production in Japan of the fieldgun, 75mm, M 30 (Osaka Arsenal/Schneider?) and the field howitzer, 105mm, M 29 (Osaka Arsenal).

But the numbers of the guns acquired were not large enough to re-equip the entire army. The rebuilt field and mountain guns therefore continued in use.

During the middle thirties the Japanese also introduced guns of their own design for use alongside their existing equipment and for replacement of the old M 08 models of the horse artillery:

fieldgun, 75mm, M 35 (Osaka Arsenal)
mountain gun, 75mm, M 34 (Osaka Arsenal)

After World War II the Japanese forces, too, received US material, chiefly the field howitzer, 105mm, M-2A1 (Ordnance Department).

Latvia

Like the two other Baltic countries Latvia had no arms industry of her own.

Unlike the armies of the two neighbours, the Latvian army was uniformly equipped with British guns. It only had fieldguns:

fieldgun, 76.2mm, M 04 (Vickers)
fieldgun, 83.8mm, Mk IV (Vickers)

Lithuania

This country, too, had no arms industry.

Lithuania's army at first used captured Russian guns. Later, German and French supplies were added:

fieldgun, 75mm, M 97 (Schneider)
fieldgun, 76.2mm, M 02 (Putilow)
field howitzer, 105mm, M 16 (Rheinmetall)

Montenegro

In the total absence of industry, the country was not in a position to develop or produce guns of its own.

Before World War I Montenegro had mainly obsolete guns, such as the fieldgun, 77mm, M 96 (Krupp) and old mountain guns, manufactured by Krupp, as well as a number of guns bought in Italy in 1905. During the war mainly French fieldguns were acquired, some of which also had French crews.

The Netherlands

A Dutch arms industry was not established until after World War I, when the firm of Siderius was founded by middlemen of Krupp's. But it confined itself to modernizing the existing fieldguns.

Before World War I the Dutch army was equipped with German guns only, the fieldgun, 75mm, M 03 (Krupp) and the field howitzer, 120mm, M 14 (Krupp).

After World War I the existing light fieldguns were modernized:

fieldgun, 75mm, M 02/04 vd OM (Siderius/Krupp)
fieldgun, 75mm, M 02/04 vd (Siderius/Krupp)
fieldgun, 75mm, M 02/04 vd NM (Siderius/Krupp)

These fieldguns constituted the largest part of the gun park of the Dutch field artillery. In addition, the Dutch bought the Swedish field howitzer, 105mm (Bofors) for their colonial army only.

Immediately before the beginning of World War II the field howitzer, 105mm, M 39 (Krupp) was ordered in Germany, but delivery was not completed.

After World War II the Dutch army, too, was at first equipped with the British gun howitzer 87.6mm, Mk II (Royal Ordnance) and later uniformly equipped with the US field howitzer, 105mm, M-2A1 (Ordnance Department).

Norway

Norwegian-built guns were first introduced between the two world wars. The firm of Kongsberg produced several models for the Norwegian army.

Before World War I the Norwegian artillery units were uniformly equipped with German guns:

fieldgun, 75mm, M 01 (Ehrhardt)
field howitzer, 120mm, M 09 (Rheinmetall)
mountain gun, 75mm, M 10 (Ehrhardt)

After World War I, the existing gun park was augmented by Norwegian designs:

field howitzer, 120mm, M 32 (Kongsberg)
mountain gun, 75mm, M 27 (Kongsberg)

After World War II, the Norwegian army used, in addition to abandoned German guns, mainly US arms like most Western armies. At present, the following field guns are in use:

field howitzer, 105mm, M 18 (Rheinmetall)
field howitzer, 105mm, M-2A1 (Ordnance Department)

Poland

Because of ther turbulent history, Poland, too, had no industrial capacity for supplying her own fieldguns.

After the achievement of independence the Polish army was, as a result of the country's political alignment, equipped mainly with French guns. These were augmented by remaining stocks of the Austro-Hungarian army as well as by Russian guns, which were allocated to the Horse Artillery. Up to the beginning of World War II the Polish army was equipped with the following guns:

fieldgun, 75mm, M 97/17 (Schneider)
fieldgun, 75mm, M 02/26 (state, Soviet Union)
field howitzer, 100mm, M 14 (Škoda)
field howitzer, 105mm, M 16 (Rheinmetall)
mountain gun, 65mm, M 06 (Schneider)
mountain gun, 75mm, M 19 (Schneider)
mountain howitzer, 100mm, M 16 (Škoda)

In addition, small numbers of the British field howitzer, 114mm, M 10 (Vickers) were used.

After World War II the Polish People's Army was equipped exclusively with Soviet guns:

fieldgun, 76.2mm, M 42 (state, Soviet Union)
fieldgun, 85mm, M 45 (state, Soviet Union)
fieldgun, 85mm, M 45 (D-48) (state, Soviet Union)
field howitzer, 122mm, M 38 (state, Soviet Union)

The modern Soviet gun howitzer, 122mm, M 63 (state) has not yet been supplied to Polish units.

Portugal

Like many other small powers, Portugal has no arms industry of her own and depends on imported material for the equipment of her army. Tight economic conditions often permitted the purchase of obsolescent guns only. In the following case this is apparent from the difference between the manufacturers' designation of the model and the Portuguese one. Only during the period before World War I was modern equipment bought, since it had become necessary to replace obsolete guns with recoil carriage.

Two guns from the time before World War I were used up to the end of World War II.

fieldgun, 75mm, M 03 (Schneider)
mountain gun, 70mm, M 07 (Schneider)

After World War I further French and British guns were added:

fieldgun, 75mm, M 97 (Schneider) (Portuguese designation M 917)
field howitzer, 114mm, M 10 (Vickers) (Portuguese designation M 917)

Many of these guns remained in use even after World War II. The latest addition was the US field howitzer, 105mm, M-2A1 (Ordnance Department).

Rumania

As an agrarian country, Rumania, like all the other Balkan countries, had no heavy industry worthy of note and was therefore not in a position to produce guns for the equipment of her army.

Before World War I the Rumanian army had, in addition to old recoil carriage guns, uniform equipment in the fieldgun, 75mm, M 03 (Krupp) and the field howitzer, 105mm, M 10 (Schneider).

This situation changed radically during World War I. The following fieldguns were added to the existing gun park through purchases abroad and from captured material:

fieldgun, 75mm, M 97 (Schneider)
fieldgun, 76.2mm, M 02 (Putilow)
fieldgun, 76.5mm, M 17 (Škoda)
field howitzer, 100mm, M 14 (Škoda)
field howitzer, 105mm, M 12/16 (Krupp)
field howitzer, 114mm, M 10 (Vickers)
mountain gun, 75mm, M 15 (Škoda)

Towards the end of the twenties an attempt was made to standardize the gun park. But the efforts failed because of lack of funds. The only result was the addition of yet another model to the already existing ones, the fieldgun, 75mm, M 28 (Škoda).

Immediately before the outbreak of World War II a second attempt at standardization of the fieldgun park was made. In addition to the existing light fieldgun by Škoda, a field howitzer, 105mm (Schneider) was to be introduced as a standard weapon. The outbreak of the war, however, prevented the execution of this plan and the gun could not be delivered. Only the mountain artillery obtained new weapons, gun and howitzer on a standard carriage:

mountain gun, 75mm, M 39 (Škoda)
mountain howitzer, 105mm, M 39 (Škoda)

After World War II, the Rumanian People's Army, like the other armies of the Eastern Bloc, was equipped with Soviet guns:

fieldgun, 76.2mm, M 42 (state, Soviet Union)
fieldgun, 85mm, M 45 (state, Soviet Union)
field howitzer, 122mm, M 38 (state, Soviet Union)

In Rumania, too, the new Soviet gun howitzer, 122mm, M 63 (state) has not yet been seen.

Russia/Soviet Union

Before World War I, guns for the Russian army were built by Putilow and Obuchow. Insufficient production capacity and lack of performance of some of these guns, meant that equipment, mainly field howitzers and mountain guns, had to be imported from Schneider and Krupp, who were leading in this field at the time. In some cases models of the same performance were ordered side by side from both firms. After the 1917 revolution gun design and production were taken over by state-owned factories.

As early as the beginning of the century Russia introduced recoil barrel guns:

fieldgun, 76.2mm, M 00 (Putilow)
field howitzer, 127mm, M 97 (?/Great Britain)

The field howitzers, however, were few in number. These guns were used in the Russo-Japanese War. Experiences of this war led to the improvement of the fieldgun and the introduction of a new light field howitzer:

fieldgun, 76.2mm, M 02 (Putilow)
field howitzer, 122mm, M 04 (Putilow/Obuchow)

Since the performance of the light field howitzer did not come up to expectations, foreign field howitzers were acquired soon afterwards:

field howitzer, 122mm, M 09 (Krupp)
field howitzer, 122mm, M 10 (Schneider)

Several models of mountainguns were also used:

mountain gun, 76.2mm, M 04 (Obuchow)
mountain gun, 76.2mm, M 09 (Schneider)
mountain howitzer, 105mm, M 09 (Schneider)

During World War I the two light fieldguns and the two light field howitzers obtained from abroad together with the imported mountain guns constituted the equipment of the Russian field artillery.

After World War I the Russians were in no position to design their own guns because of the revolution and the civil war following it. During this period, only captured guns, such as the field howitzer, 114mm, M 10 (Vickers) were added to the existing armoury of the artillery units.

Not until the beginning of the thirties did it become possible to consider the improvement of the old guns. Above all their range was to be increased. The Russians succeeded in this partly by rebuilding them, using new carriages, and increasing the muzzle velocity. The results were the fieldgun, 76.2mm, M 02/30 (state/Putilow), the field howitzer, 122mm,

M 10/30 (state/Schneider) and the field howitzer, 122mm, M 09/37 (state/Krupp).

In addition, however, work was already proceeding on the development of new models, particularly of light fieldguns:

fieldgun, 76.2mm, M 33 (state, Soviet Union)
fieldgun, 76.2mm, M 36 (state, Soviet Union)
fieldgun, 76.2mm, M 39 (state, Soviet Union)
field howitzer, 122mm, M 38 (state, Soviet Union)
mountain gun, 76.2mm, M 38 (state, Soviet Union)

With these guns the Red Army entered World War II. Almost all the older models were lost during the reverses suffered at the beginning of the war. The following light fieldguns were added in its course:

fieldgun, 76.2mm, M 41 (state, Soviet Union)
fieldgun, 76.2mm, M 39/42 (state, Soviet Union)
fieldgun, 76.2mm, M 42 (state, Soviet Union)

Of these three models, however, only the last-named was built in large numbers.

During the latter half of the war, the Russians began to increase the fire power of the light field gun by increasing its calibre. The results were the fieldgun, 85mm, M 43 (state, Soviet Union) and the fieldgun, 85mm, M 45 (state, Soviet Union).

The fieldguns were increasingly called upon to reinforce anti-tank defence, which ultimately became their main task.

After World War II the Soviet artillery retained only three light fieldguns, the fieldgun, 76.2mm, M 42, the fieldgun, 85mm, M 45, and the field howitzer, 122mm, M 38. Since even now the light fieldguns are used mainly against armoured vehicles, they are usually called anti-tank guns in the West.

In the course of improvement of performance the light fieldgun, 76.2mm, M 42 was scrapped, and is at present found only in some minor Eastern Bloc countries. The light fieldgun, 85mm, M 45 and the light field howitzer, 122mm, M 38, used as main equipment during the period after the war, have now also been relegated to reserve units in the Soviet Union. Two new guns have recently been introduced:

fieldgun, 85mm, M 45 (D-48) (state, Soviet Union)
gun howitzer, 122mm, M 63 (state, Soviet Union)

The gun howitzer on a universal carriage must be considered the most modern fieldgun at present. It is expected to become the standard gun of the divisional artillery. It can also be used as an anti-tank gun because of its 360° traversing range, low trajectory, and high muzzle velocity.

Serbia/Yugoslavia

Up to the end of World War II Serbia did not manufacture her own arms. As an agrarian country Serbia, which later became Yugoslavia, lacked the heavy industry that could have served as a basis for the production of guns. Only during the fifties was an efficient steel industry established in the course of the country's general industrialization, and very soon the production of Yugoslav guns began.

Before World War I, Serbia's political alignment naturally affected the choice of her weapons. All the guns bought during this period came from France:

fieldgun, 75mm, M 06 (Schneider)
fieldgun, 75mm, M 12 (Schneider)
field howitzer, 120mm, M 97 (Schneider)
field howitzer, 120mm, M 11 (Schneider)
mountain gun, 70mm, M 07 (Schneider)

Owing to the events during World War I a large proportion of this stock was lost.

After World War I, stocks were replenished mainly with captured or surrendered guns of the Austro-Hungarian army. The models mainly used were:

fieldgun, 76.5mm, M 05/08 (Škoda)
field howitzer, 100mm, M 14 (Škoda)
mountain gun, 75mm, M 15 (Škoda)
mountain howitzer, 100mm, M 16 (Škoda)

During the inter-war years Yugoslavia was one of the first countries to renew her stocks of fieldguns. It was planned to equip the entire field and mountain artillery with standard weapons, and both the fieldguns (light fieldgun and light field howitzer) and the mountain guns (mountain gun and mountain howitzer) were to be mounted on a standard carriage. In contrast with the period before World War I the guns were no longer bought in France, but in Czechoslovakia. The following guns were introduced:

fieldgun, 75mm, M 28 (Škoda)
field howitzer, 100mm, M 28 (Škoda)
mountain gun, 75mm, M 28 (Škoda)
mountain howitzer, 90mm, M 28 (Škoda)

Because of financial stringency, however, the aim of standardizing the equipment could not be achieved, so that all the field and mountain guns introduced before World War I and the Austro-Hungarian models had to be used together with the modern M 28 guns until the start of World War II.

33

After World War II, the light artillery used mainly abandoned German guns, such as the field howitzer, 105mm, M 18M (Rheinmetall/Krupp) and US guns supplied later—the field howitzer, 105mm, M-2A1 (Ordnance Department).

The establishment of Yugoslav heavy industry soon created facilities for the development of Yugoslav guns, of which there are two models:

field howitzer, 105mm, M 56 (state, Yugoslavia)
mountain gun, 75mm, M 48 (state, Yugoslavia)

These two models together with the German and US light field howitzers from World War II are the current light fieldguns of the Yugoslav People's Army.

Soviet Union

See Russia/Soviet Union.

Spain

In spite of her size, Spain never had the heavy industry that could have formed the basis for the design of her own guns. Whether the two recently noted light fieldguns were built in Spain is unknown.

Before World War I Spain, like Portugal—her neighbour on the Iberian peninsula, was equipped almost exclusively with French guns. These were the fieldgun, 75mm, M 06 (Schneider) and the mountain gun, 70mm, M 08 (Schneider).

After World War I French and British guns were bought:

field howitzer, 105mm, M 22 (Vickers)
mountain howitzer, 105mm, M 19 (Schneider)

With these four types of gun the two sides fought each other in the Spanish Civil War 1936/9. Owing to the intervention of foreign elements many models of British, German, French, Italian, and Soviet manufacture entered the country. After the end of the Civil War, the German field howitzer, 105mm, M 18 (Rheinmetall) was adopted as the standard gun.

During the fifties the United States supplied large numbers of the field howitzer, 105mm, M-2A1 (Ordnance Department) also to Spain. In addition, the Spanish army has two other fieldguns of unknown manufacture:

fieldgun, 75mm (Spain?)
field howitzer, 105mm (Spain?)

The most modern gun is the Italian mountain howitzer, 105mm, M 56 (OTO Melara) bought within the last few years.

Sweden

In the firm of Bofors Sweden has one of the world's best-known arms factories. Although this company has manufactured guns since 1883, foreign fieldguns were imported up to World war I.

Before World War I the Swedish army was equipped with two models:

fieldgun, 75mm, M 02 (Krupp)
field howitzer, 105mm, M 10 (Bofors)

Only in the early thirties did the modernization of the existing gun park begin; some of the light fieldguns were rebuilt, eg the fieldgun, 75mm, M 02/33 (Bofors/Krupp).

But in addition, the old models continued in use up to the beginning of World War II.

Not until the beginning of that war was the gun park radically renewed. A light fieldgun and a light field howitzer were introduced on a standard carriage. But since existing production capacity was insufficient to equip the entire army a German light field howitzer, the field howitzer M 39, had to be imported. The equipment therefore consisted of the following guns:

fieldgun, 75mm, M 40 (Bofors)
field howitzer, 105mm, M 40 (Bofors)
field howitzer, 105mm, M 18 (Rheinmetall)

After World War II only the two modern light field howitzers were retained. The new field howitzer, 105mm, 4140 (Bofors) was added. With this gun Sweden was the first country to introduce a light field howitzer on a universal carriage in large numbers. Currently a new fieldgun with auxiliary drive is being developed for the Swedish army.

Switzerland

Before World War I Switzerland had no arms industry worth mentioning. Only as a result of development after that war did the Swiss begin to rebuild, as well as to manufacture their guns. The firm of Sulzer was engaged mainly in rebuilding. Even after World War II production was confined to the licensed production of guns designed by Bofors.

All the guns introduced before World War I were made by Krupp:

fieldgun, 75mm, M 03 (Krupp)
field howitzer, 120mm, M 12 (Krupp)
mountain gun, 75mm, M 06 (Krupp)

After World War I the existing light fieldgun was rebuilt in Switzerland ie the fieldgun, 75mm, M 03/22 (Sulzer/Krupp).

Later the obsolete mountain gun was replaced by a new one—the mountain gun, 75mm, M 33 (state/Bofors).

During the war some barrels of the fieldgun were put on a new split trail carriage—the fieldgun, 75mm, M 03/40 (Krupp) and used with the light brigades.

After World War II, the obsolete fieldguns, too, were replaced by a new model, the field howitzer, 105mm, M 46 (state).

The mountain guns have now been scrapped.

Turkey

Although a comparatively large country, Turkey lacked an industrial base that could have sustained the design and production of her own guns.

At the beginning of World War I the Turkish army was equipped mainly with Krupp guns. In addition to obsolete recoil carriage guns (field howitzer, 120mm, M 92 (Krupp) the main armament consisted of:

fieldgun, 75mm, M 03 (Krupp)
mountain gun, 75mm, M 04 (Krupp)
mountain gun, 75mm (MD) (Schneider)

A modern field howitzer ordered from Austria-Hungary was delivered only after Turkey had entered the war on the side of the Central Powers. This was the field howitzer, 105mm, M 12 (Škoda).

In the course of assistance with war supplies during World War I mainly German guns reached Turkey. They were:

fieldgun, 77mm, M 96 (new type (Krupp/Ehrhardt)
fieldgun, 77mm, M 16 (Rheinmetall)
field howitzer, 105mm, M 98/09 (Krupp)
field howitzer, 105mm, M 16 (Rheinmetall)
mountain gun, 75mm, M 14 (Rheinmetall)
mountain gun, 77mm, M 15 (Rheinmetall)
mountain howitzer, 105mm (Krupp)

The mountain gun, 75mm, M 15 (Škoda) was supplied by Austria-Hungary.

Between the two world wars the bulk of the equipment was made up of three existing types:

fieldgun, 77mm, M 16 (Rheinmetall)
field howitzer, 105mm, M 16 (Rheinmetall)
mountain gun, 75mm, M 15 (Škoda)

Modern mountain guns were introduced:

mountain gun, 75mm, M 30 (Bofors)
mountain howitzer, 100mm, M 16/19 (Škoda)

After World War II the US field howitzer, 105mm, M-2A1 (Ordnance Department) became the standard equipment of the Turkish artillery. In addition, small numbers of a mountain howitzer (Bofors?) are said to exist.

USA

Although the US has always had important heavy industries, the country's output of guns was for a long time insignificant. Of the well-known steel firms only the Bethlehem Steel Corporation was active in this field. In contrast to many other countries, however, in the US it was the army administration itself, the Ordance Department, which established itself with its own designs as early as the beginning of the century. But because of the strategic situation and the political outlook production lagged far behind requirements, so that the US army had to enter both world wars more or less unprepared.

At the beginning of the century the artillery was modernized in the US as in all other countries. After protracted trials the fieldgun, 76.2mm, M 02 (Ordnance Department) was introduced. For the mountain artillery the mountain howitzer, 75mm, M 01 (Vickers) was bought in Great Britain. Shortly before the outbreak of war the mountain howitzer, 76.2mm, M 11 (?/USA) was introduced.

The fieldgun, 96.5mm, M 07 (Ordnance Department) and the field howitzer, 119mm, M 08 (Ordnance Department) were only produced in small numbers for use by the army.

As early as 1912 work began on a new fieldgun, the fieldgun, 75mm, M 16 (Ordnance Department) which was, however, delivered only shortly before the US entered World War I.

At the same time production started in the US of a light fieldgun according to British plans and of a light field howitzer:

fieldgun, 75mm, M 17 (Bethlehem)
field howitzer, 96.5mm, M 17 (Ordnance Department)

After the US entered World War I, however, it very soon became obvious that neither the existing stocks nor the production capacity were sufficient for the equipment of the army on a war footing. The American troops in Europe were therefore equipped mainly with the French fieldgun, 75mm, M 97 (Schneider). The extent of this new equipment is shown by the fact that at the end of World War I the American troops had more than twice as many French as American fieldguns.

After the end of World War I no expense was spared in the development of modern fieldguns. But the effort soon slackened and except for a few experimental types no new guns were built. The army therefore retained the French light fieldgun, 75mm, M 97 and the current American guns, the light fieldgun, 75mm, M 16 and M 17, the light field howitzer, 96.5mm, M 17, and the old mountain howitzer, 75mm, M 01 as its armament.

The necessary re-equipment was not carried out until later. In addition to a mountain howitzer a motorized carriage was introduced for the French light fieldgun: this produced the fieldgun, 75mm, M 97A4 (Ordnance Department/Schneider) and the mountain howitzer, 75mm, M I (Ordnance Department).

In addition, many experimental guns were developed and prepared for mass production in the event of war.

At the beginning of World War II mass production started as planned; but because of European experiences production of the light fieldgun soon stopped. The gun park consisted mainly of the following types:

fieldgun, 75mm, M-2A2 (Ordnance Department)
field howitzer, 75mm, M-3A1 (Ordnance Department)
field howitzer, 105mm, M-2A1 (Ordnance Department)
mountain howitzer, 75mm, M-1A1 (Ordnance Department)

In addition, however, the old guns motorized carriages remained in use until after the outbreak of war.

During the war the industrial stength of the USA became evident. Introduction of new, improved models was abandoned in favour of mass production. The light field howitzer, 105mm, M-2A1 thus became the standard gun of the US field artillery. The mountain howitzer, 75mm, M-1A1 was used by the Marines and airborne units. The only new gun introduced was a light field howitzer, the field howitzer, 105mm, M-3A1 (Ordnance Department) for use by airborne troops and in jungle warfare; it was mounted on the carriage of the already scrapped field howitzer, 75mm, M-3A1.

After World War II the US army retained only the field howitzer, 105mm, M-2A1 and the mountain howitzer, 75mm, M-1A1. After minor modifications they were redesignated M-101 and M-116. Only at the beginning of the sixties was a new fieldgun, the field howitzer, 105mm, M-102 (Ordnance Department) introduced, but for the time being for airborne units only. At the moment trials are being conducted of numerous new models, some with auxiliary drive, in the USA. Among others, the principle of the counter-recoil gun is being considered.

Yugoslavia
See Serbia/Yugoslavia.

Tables of Types

Key to the technical terms used

The **sequence of production** in the manufacturing countries is arranged according to calibre groups and within these in the alphabetical order of the manufacturers. Here, the sequence is based on the models or on their year of manufacture.

The **designation** corresponds to the most recent version in the sequence: type of gun, calibre, model (manufacturer's designation).

the **manufacturer's name** (unless it is a government agency) may, in some countries, include the name of the designer.

The **weight in firing position** is generally the weight as constructed or average weight. Deviations have often occurred because the guns were built by various manufacturers.

The **weight of the projectile**, unless otherwise stated, is that of the high-explosive shell.

The **muzzle velocity** indicates the maximum velocity.

The **maximum range** indicates the technically possible range. With some of the pre-1914 models it could be achieved only with the trail dug in. Some sighting mechanisms, too, prevented the utilization of the maximum range.

The description of the **carriage** distinguishes only between box trail and split-trail, and, where applicable, pole-trail carriage.

Mode of traction: Almost all the old guns were designed only for horse traction. Modifications for motor traction were, however, often introduced later, or motorization was improvised.

'Introduced into'; generally only introduction by European countries and the non-European Great Powers (USA and Japan) is listed. Captured guns used only during wartime are not listed.

Rebuilt guns are listed under the country of rebuilding.

Later modifications, new types of ammunition etc are indicated by annotations after an oblique stroke in the relevant columns.

Manufacturers

Austria-Hungary/Austria
Artillerie-Zeugs-Fabrik (AZF)
Gebröder Böhler
Škoda (Škodawerke AG)
state

Belgium
Cockerill (Société anonyme John Cockerill)
FRC (Fonderie Royale des Canons)

Czechoslovakia
Škodawerke AG
state (formerly Škodawerke AG)

Finland
Tampella Ab

France
Schneider & Cie
St Chamond (Compagnie des Forges et Aciéries de la Marine et d'Homécourt in Saint Chamond)
state (Arsenals in Bourges, Puteaux etc)
DTAT (Direction Techniques des Armements Terrestres)
Designers: de Bange, Baquet, Deport.

Germany
Ehrhardt (Rheinische Metallwaren—und Maschinenfabrik) later known as: Rheinmetall-Borsig AG, now:
Rheinmetall GmbH
Krupp (Fried. Krupp AG)

Great Britain
Armstrong Whitworth & Co
Vickers (Vickers, Sons and Maxim, later—Vickers-Armstrong Ltd)
state (Woolwich Arsenal, Coventry Ordnance Works, Armaments Research and Development Establishment)

Hungary
MÁVAG
state

Italy
Ansaldo
Arsenal Turin
OTO (Oderno Terni Orlando) now:
OTO Melara
Vickers-Terni

Japan
state (Osaka Arsenal)
Designers: General Arisaka

The Netherlands
Siderius AG

Norway
AB Kongsberg

Russia/Soviet Union
Obuchow
Putilow
state

Spain
Unknown

Sweden
Aktiebolaget Bofors

Switzerland
Gebrüder Sulzer
state

United States of America
Bethlehem Steel Corporation
state (Ordnance Department, Rock Island Arsenal)

Yugoslavia
state

Country of origin: Austria-Hungary
Manufacturer: Böhler
Designation: 8cm M 18 Fieldgun
Calibre (mm): 76.5
Length of barrel (x calibre): 30
Depression/Elevation (°): -10 +45
Traverse (°): 8
Weight in firing position (kg): 1,330
Weight of projectile (kg): 8.0
Muzzle velocity (m/sec): 500
Maximum range (km): 10.5 *
Gun carriage: Box trail
Gun shield: Yes
Mode of traction: Horse and motor drawn
Introduced into: Austria-Hungary, Austria, Hungary

Remarks:

Introduced simultaneously with the 76.5mm Fieldgun M 17 (Skoda).
Capable of firing the same ammunition as the two older 76.5mm
Fieldguns, M 05 and M 05/08 (Škoda).

* In Austria with the M 32 shell 12.5km.

Country of origin: Austria-Hungary
Manufacturer: state/Škoda
Designation: 8cm Fieldgun M 5
Calibre (mm): 76.5
Length of barrel (x calibre): 30
Depression/Elevation (^{O}): -7.5 +18
Traverse (^{O}): 8
Weight in firing position (kg): 1,020
Weight of projectile (kg): 6.68
Muzzle velocity (m/sec): 500
Maximum range (km): 7.0
Gun carriage: Box trail
Gun shield: Yes
Mode of traction: Horse drawn
Introduced into: Austria-Hungary

Remarks:
Replaced the older recoil carriage guns. The main weapon of the fieldgun
regiments at the beginning of World War I. Replaced by the 76.5mm
Fieldguns, M 17 (Škoda) and M 18 (Böhler). Barrel of wrought bronze.

Country of origin: Austria-Hungary
Manufacturer: state/Škoda
Designation: 8cm M 5/8 Fieldgun
Calibre (mm): 76.5
Length of barrel (x calibre): 30
Depression/Elevation (°): -5 +23
Traverse (°): 8
Weight in firing position (kg): 1,020
Weight of projectile (kg): 6.68
Muzzle velocity (m/sec): 500
Maximum range (km): 7.0*
Gun carriage: Box trail
Gun shield: Yes
Mode of traction: Horse drawn
Introduced into: Austria-Hungary, Yugoslavia, Hungary, Czechoslovakia

Remarks:

Dismantles into three parts and transported in narrow gauge carts in mountainous areas. Improved 76.5mm Fieldgun M 05 (state/Škoda). Also adapted for use in the mountains. At the beginning of World War I there were at the most two batteries of the Fieldgun regiments equipped with this gun. Barrel of wrought bronze.

*In Czechoslovakia with M 30 shells and entrenched trail spade 9.3km.

Fieldgun 75mm M 11   Austria-Hungary

Country of origin: Austria-Hungary
Manufacturer: Škoda
Designation: 7.5cm Fieldgun M 12
Calibre (mm): 75
Length of barrel (x calibre): 29
Depression/Elevation ($^{\circ}$): -8 +16
Traverse ($^{\circ}$): 7
Weight in firing position (kg): 940
Weight of projectile (kg): 6.5
Muzzle velocity (m/sec): 500
Maximum range (km): 6.0
Gun carriage: Box trail
Gun shield: Yes
Mode of traction: Horse drawn
Introduced into: China, Austria-Hungary, Czechoslovakia

Remarks:
The second appointed delivery of 24 of these guns to China was confiscated at the outbreak of World War I and used by the Austro-Hungarian army on the Russian front. the guns remained in use by the Czechoslovakian Artillery Reserve until 1938.

Country of origin: Austria-Hungary
Manufacturer: Škoda
Designation: 8cm M 17 Fieldgun
Calibre (mm): 76.5
Length of barrel (x calibre): 30
Depression/Elevation (^{O}): -10 +45
Traverse (^{O}): 8
Weight in firing position (kg): 1,386
Weight of projectile (kg): 8.0
Muzzle velocity (m/sec): 500
Maximum range (km): 9.9*
Gun carriage: Box trail
Gun shield: Yes
Mode of traction: Horse and motor drawn
Introduced into: Austria-Hungary, Austria, Czechoslovakia, Rumania

Remarks:

The gun was intended, together with the 76.5mm Fieldgun M 18 (Böhler), as a replacement for the older 76.5mm Fieldguns M 05 and M 05/08 (state/Škoda). Apart from the new ammunition, the gun was also capable of firing the ammunition of both the older guns. Continued to be built in Czechoslovakia until 1938.

*In Austria with M 32 shell 12km.

Country of origin: Austria-Hungary
Manufacturer: state
Designation: 10cm Field Howitzer M 99
Calibre (mm): 104
Length of barrel (x calibre): 13
Depression/Elevation (°): -10 +42.5
Traverse (°): —
Weight in firing position (kg): 1,000
Weight of projectile (kg): 14.7
Muzzle velocity (m/sec): 313
Maximum range (km): 6.0
Gun carriage: Box trail
Gun shield: No
Mode of traction: Horse drawn
Introduced into: Austria-Hungary

Remarks:
Recoil gun with trail spade and rope brake. Similar to the 104mm Field
Howitzer M 99 (narrow gauge) for mountain use. The main equipment of
the Field Howitzer regiments at the start of World War I. During the war
there were two batteries, thus equipped, to each Division of the Infantry.
Barrel of wrought bronze.

Country of origin: Austria-Hungary
Manufacturer: Škoda
Designation: 10.5cm M 14/T Field Howitzer*
Calibre (mm): 105
Length of barrel (x calibre): 18
Depression/Elevation (º): -8 +70
Traverse (º): 6
Weight in firing position (kg): 1,397
Weight of projectile (kg): 16.0
Muzzle velocity (m/sec): 350
Maximum range (km): 7.75
Gun carriage: Box trail
Gun shield: Yes
Mode of traction: Horse drawn
Introduced into: Turkey, Austria-Hungary

Remarks:

Ordered by Turkey in 1913. In the autumn of 1914, before Turkey entered
the war, fifty guns were confiscated by Austria-Hungary, and used in their
own army.

*Austro-Hungarian designation.

Country of origin: Austria-Hungary
Manufacturer: Škoda
Designation: 10.4cm M 14 China-Field Howitzer *
Calibre (mm): 104
Length of barrel (x calibre): 18
Depression/Elevation (^{o}): -8 +45
Traverse (^{o}): 7
Weight in firing position (kg): 1,250
Weight of projectile (kg): 14.7
Muzzle velocity (m/sec): 310
Maximum range (km): 6.3
Gun carriage: Box trail
Gun shield: Yes
Mode of traction: Horse drawn
Introduced into: China, Austria-Hungary

Remarks:
A number of these guns, destined for China, were taken over by
Austria-Hungary at the beginning of the war, and used in their own army.

*Austro-Hungarian designation.

Field Howitzer 100mm M 14

Country of origin: Austria-Hungary
Manufacturer: Škoda
Designation: 10cm Field Howitzer M 14
Calibre (mm): 100
Length of barrel (x calibre): 19.3
Depression/Elevation (°): -8 +48
Traverse (°): 5
Weight in firing position (kg): 1,420
Weight of projectile (kg): 11.5/16.0
Muzzle velocity (m/sec): 422/341*
Maximum range (km): 8.0*
Gun carriage: Box trail
Gun shield: Yes
Mode of traction: Horse and motor drawn
Introduced into: Austria-Hungary, Succession States, Italy, Poland, Rumania

Remarks:

Barrel similar to 100mm Mountain Howitzer M 16. Dismantles into 3 parts for mountain transportation. Early in 1915, replaced the 104mm Field Howitzer M 99 (state). For production reasons, on testing selected field howitzers, it was decided that the gun should be built instead of that produced by Böhler. The gun was partly rebuilt for motor traction in Italy after World War I.

*In Austria with the M 32 shell 397 m/sec and 9.4km. In Italy with the M 32 shell 13.65kg, 407 m/sec and 9.3km.

Country of origin: Austria-Hungary
Manufacturer: state
Designation: 7cm M 99 Mountain Gun
Calibre (mm): 72.5
Length of barrel (x calibre): 13.8
Depression/elevation (°): -10 +26
Traverse (°): —
Weight in firing position (kg): 318
Weight of projectile (kg): 4.68/4.85
Muzzle velocity (m/sec): 304/297
Maximum range (km): 5.2
Gun carriage: Box trail
Gun shield: No
Mode of traction: By pack animals
Introduced into: Austria-Hungary

Remarks:
Dismantles into three parts. Recoil carriage gun with trail spade. The main equipment of the mountain batteries at the beginning of World War I. Barrel of wrought bronze.

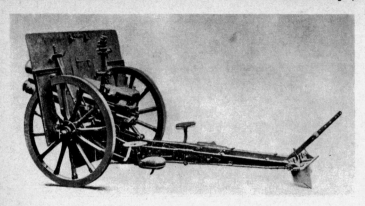

Country of origin: Austria-Hungary
Manufacturer: Artillerie-Zeugs-Fabrik (AZF)
Designation: 7cm Mountain Gun M 8
Calibre (mm): 72.5
Length of barrel (x calibre): —
Depression/Elevation (^{o}): -12 +34
Traverse (^{o}): 8
Weight in firing position (kg): 402
Weight of projectile (kg): 4.8
Muzzle velocity (m/sec): 310
Maximum range (km): 5.3
Gun carriage: Box trail
Gun shield: Yes
Mode of traction: Horse drawn, pack animals
Introduced into: Austria-Hungary

Remarks:

Dismantles into four parts. Introduced in small numbers together with the
similar 72.5mm Mountain Gun M 09 (Škoda). At the beginning of World
War I there were, altogether, only 20 batteries using the two models. Both
models fired the ammunition of the 72.5mm Mountain Gun M 99 (state).
Barrel of wrought bronze.

Country of origin: Austria-Hungary
Manufacturer: Škoda
Designation: 7cm Mountain Gun M 9
Calibre (mm): 72.5
Length of barrel (x calibre): —
Depression/Elevation (^{O}): -9 +35
Traverse (^{O}): —
Weight in firing position (kg): 456
Weight of projectile (kg): 4.8
Muzzle velocity (m/sec): 310
Maximum range (km): 5.3
Gun carriage: Box trail
Gun shield: Yes
Mode of traction: Horse drawn, pack animals
Introduced into: Austria-Hungary

Remarks:

Dismantles into five parts. Similar to the 72.5mm Mountain Gun M 08
(Artillerie-Zeugs-Fabrik). Barrel of wrought bronze.

Country of origin: Austria-Hungary
Manufacturer: Škoda
Designation: 7.5cm M 14 China-Mountain Gun
Calibre (mm): 75
Length of barrel (x calibre): 15
Depression/Elevation (°): -10 +30
Traverse (°): 7
Weight in firing position (kg): 470
Weight of projectile (kg): 5.3
Muzzle velocity (m/sec): 300
Maximum range (km): 5.6
Gun carriage: Box trail
Gun shield: Yes
Mode of traction: By pack animals
Introduced into: China, Austria-Hungary

Remarks:

Dismantles into five parts. Similar guns delivered to Uruguay, Ecuador and
Costa Rica. Ordered by China. After the beginning of World War I
Austria-Hungary took over 52 of these guns. In use from autumn 1914.

Country of origin: Austria-Hungary
Manufacturer: Škoda
Designation: 7.5cm M 15 Mountain Gun
Calibre (mm): 75
Length of barrel (x calibre): 15
Depression/Elevation (^{o}): -9 +50
Traverse (^{o}): 7
Weight in firing position (kg): 620
Weight of projectile (kg): 6.5
Muzzle velocity (m/sec): 350
Maximum range (km): 7.0 *
Gun carriage: Box trail
Gun shield: Yes
Mode of traction: Horse drawn, pack animals
Introduced into: Austria-Hungary, Succession States, Bulgaria, Germany,
   Italy, Rumania, Turkey

Remarks:

Dismantles into six parts. From 1915 replaced the obsolete 72.5mm
Mountain Gun M 99 (state). Used by the Austro-Hungarian Succession
States, and other states until the beginning of World War II. Also used to a
certain extent as an infantry gun.

* In Austria with the M 32 shell 8.1km.

Country of origin: Austria-Hungary
Manufacturer: Škoda
Designation: 10cm M 8 Mountain Howitzer
Calibre (mm): 104
Length of barrel (x calibre): 14.7
Depression/Elevation (^{o}): -8 +42
Traverse (^{o}): 5
Weight in firing position (kg): 1,233
Weight of projectile (kg): 14.7
Muzzle velocity (m/sec): 300
Maximum range (km): 6.0
Gun carriage: Box trail
Gun shield: Yes
Mode of traction: Horse drawn
Introduced into: Austria-Hungary

Remarks:

Two loads tandem drawn. Replaced by the 100mm Mountain Howitzer
M 16 (Škoda). Like the similar 104mm Mountain Howitzer M 10 (Škoda),
this gun fired the ammunition of the 104mm Field Howitzer M 99 (state).
Barrel of wrought bronze.

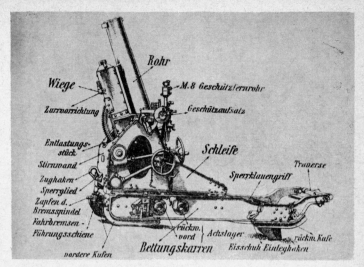

Country of origin: Austria-Hungary
Manufacturer: Škoda
Designation: 10cm M 10 Mountain Howitzer
Calibre (mm): 104
Length of barrel (x calibre): 14.7
Depression/Elevation (^{o}): -8 +42/+43 +70 *
Traverse (^{o}): 5/6 *
Weight in firing position (kg): 1,210/990 *
Weight of projectile (kg): 14.7
Muzzle velocity (m/sec): 300
Maximum range (km): 6.0
Gun carriage: Box trail/sledge
Gun shield: Yes/No
Mode of traction: Horse drawn
Introduced into: Austria-Hungary

Remarks:

Two loads tandem drawn. Replaced by the 100mm Mountain Howitzer
M 16 (Škoda). Like the similar 104mm Mountain Howitzer M 08 (Škoda),
this gun fired the ammunition of the 104mm Field Howitzer M 99 (state).
Barrel of wrought bronze.

　* In gun carriage of Mountain Howitzer M 10/in sledge gun carriage M 8
or M 8/12.

Country of origin: Austria-Hungary
Manufacturer: Škoda
Designation: 10cm M 16 Mountain Howitzer
Calibre (mm): 100
Length of barrel (x calibre): 19.3
Depression/Elevation (^{o}): -8 +70
Traverse (^{o}): 5.5
Weight in firing position (kg): 1,235
Weight of projectile (kg): 16.0 *
Muzzle velocity (m/sec): 341 *
Maximum range (km): 7.75 *
Gun carriage: Box trail
Gun shield: Yes
Mode of traction: Horse drawn
Introduced into: Austria-Hungary, Succession States, Germany, Italy, Poland, Turkey

Remarks:
Three loads tandem drawn. Developed in conjunction with the 100mm Field Howitzer M 14 (Škoda), many of the parts were interchangeable and the barrels were indentical. Delivered to Turkey as the 105mm Mountain Howitzer M 16 (T), also used in Germany with a 105mm calibre.

 * In Italy with the M 32 shell 13.37kg, 407 m/sec and 9.3km.

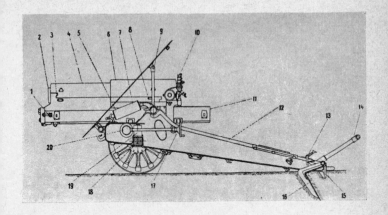

Country of origin: Austria-Hungary
Manufacturer: Škoda
Designation: 15cm M 18 Mountain Howitzer
Calibre (mm): 150
Length of barrel (x calibre): 13
Depression/Elevation (°): -5 +70
Traverse (°): 7
Weight in firing position (kg): 2,800
Weight of projectile (kg): 42.0
Muzzle velocity (m/sec): 340
Maximum range (km): 8.0
Gun carriage: Box trail
Gun shield: Yes
Mode of traction: Horse drawn
Introduced into: Austria-Hungary

Remarks:

Six loads tandem drawn. The heaviest mountain gun. By the end of World
War I only a prototype was ready.

Fieldgun 75mm (Quickfiring)

Country of origin: Belgium/Germany
Manufacturer: FRC/Krupp
Designation: 7.5cm Fieldgun (Quickfiring)
Calibre (mm): 75
Length of barrel (x calibre): 30
Depression/Elevation (°): -10 +21
Traverse (°): 6.5
Weight in firing position (kg): 1,190
Weight of projectile (kg): 6.5
Muzzle velocity (m/sec): 540
Maximum range (km): 9.9
Gun carriage: Box trail
Gun shield: Yes
Mode of traction: Horse drawn
Introduced into: Belgium

Remarks:
Adapted from the 75mm Fieldgun M 05 (Krupp).

Fieldgun 75mm GP I   Belgium

Country of origin: Belgium/Germany
Manufacturer: Cockerill/Rheinmetall
Designation: 75mm Gun Mk GP I
Calibre (mm): 75
Length of barrel (x calibre): 35
Depression/Elevation (º): -13 +42
Traverse (º): 3
Weight in firing position (kg): —
Weight of projectile (kg): 6.125
Muzzle velocity (m/sec): 579
Maximum range (km): 11.0
Gun carriage: Box trail
Gun shield: Yes
Mode of traction: Horse drawn
Introduced into: Belgium

Remarks:

Lengthened barrel of the 75mm Fieldgun M 05 (Krupp) in the gun carriage
of the 105mm Field Howitzer M 16 (Rheinmetall).

Country of origin: Belgium/Germany
Manufacturer: Cockerill/Rheinmetall)
Designation: 75mm Gun Mk GP II
Calibre (mm): 75
Length of barrel (x calibre): 37.4
Depression/Elevation (^{o}): -7 +43
Traverse (^{o}): 8
Weight in firing position (kg): 1,510
Weight of projectile (kg): 6.125
Muzzle velocity (m/sec): 579
Maximum range (km): 11.0
Gun carriage: Box trail
Gun shield: Yes
Mode of traction: Horse or motor drawn
Introduced into: Belgium

Remarks:
Modified 77mm Fieldgun M 16 (Rheinmetall).

Fieldgun 75mm GP III   Belgium

Country of origin: Belgium/Germany
Manufacturer: Cockerill/Rheinmetall)
Designation: 75mm Gun Mk GP III
Calibre (mm): 75
Length of barrel (x calibre): 37.3
Depression/Elevation (°): -8 +35
Traverse (°): 8
Weight in firing position (kg): 1,450
Weight of projectile (kg): 6.125
Muzzle velocity (m/sec): 579
Maximum range (km): 11.0
Gun carriage: Box trail
Gun shield: Yes
Mode of traction: Horse or motor drawn
Introduced into: Belgium

Remarks:

Together with the 75mm Fieldgun TR and the 75mm GP I and GP II
Fieldguns, it was the principle weapon of the Belgian Artillery until 1940.
Originally the captured German 77mm Fieldgun M 16 (Rheinmetall) which,
by inserting a barrel lining, was adapted for French 75mm ammunition.

Country of origin: Czechoslovakia
Manufacturer: Škoda
Designation: 7.5cm gun M 28
Calibre (mm): 75
Length of barrel (x calibre): 40
Depression/Elevation (^{O}): -8 +80
Traverse (^{O}): 13
Weight in firing position (kg): 1,680
Weight of projectile (kg): 7.3
Muzzle velocity (m/sec): 650
Maximum range (km): 14.0
Gun carriage: Box trail
Gun shield: Yes
Mode of traction: Horse drawn
Introduced into: Yugoslavia, Rumania

Remarks:

Standard gun carriage of the 100mm Field Howitzer M 28. One of the first designs after 1918 which were introduced as standard weapons to an army. The gun was also to be used as an anti-aircraft weapon.

Fieldgun 76.5mm M 30 (NPK)

Country of origin: Czechoslovakia
Manufacturer: Škoda
Designation: 8cm Gun M 30
Calibre (mm): 76.5
Length of barrel (x calibre): 40
Depression/Elevation (°): -8 +80
Traverse (°): 8
Weight in firing position (kg): 1,816
Weight of projectile (kg): 8.0
Muzzle velocity (m/sec): 600
Maximum range (km): 13.5
Gun carriage: Box trail
Gun shield: Yes
Mode of traction: Horse or motor drawn
Introduced into: Czechoslovakia

Remarks:

Standard carriage of the 100mm Field Howitzer M 30. Built in two forms
for either horse or motor traction. Introduced 1934. 200 guns delivered up
until 1938. The gun was also to be used as an anti-aircraft weapon. Also
adapted to fire the ammunition of the 76.5mm Fieldgun M 17 (Škoda).

Fieldgun 75mm M 35 (E3)   Czechoslovakia

Country of origin: Czechoslovakia
Manufacturer: Škoda
Designation: 7.5cm Cavalry Gun M 35
Calibre (mm): 75
Length of barrel (x calibre): 21.2
Depression/Elevation (^{0}): -9 +45
Traverse (^{0}): 50
Weight in firing position (kg): 1,040
Weight of projectile (kg): 6.3
Muzzle velocity (m/sec): 480
Maximum range (km): 10.2
Gun carriage: Split trail
Gun shield: Yes
Mode of traction: Horse drawn
Introduced into: Czechoslovakia

Remarks:
Underwent tests in 1934. Provided as equipment for the Horse Artillery
units of the Cavalry.

Country of origin: Czechoslovakia
Manufacturer: Škoda
Designation: 7.65cm Gun M 39
Calibre (mm): 76.5
Length of barrel (x calibre): 29.7
Depression/Elevation (°): -6 +45
Traverse (°): 50
Weight in firing position (kg): 1,425
Weight of projectile (kg): 7.3
Muzzle velocity (m/sec): 570
Maximum range (km): 12.0
Gun carriage: Split trail
Gun shield: Yes
Mode of traction: Horse or motor drawn
Introduced into: Czechoslovakia

Remarks:

Standard gun carriage of the 100mm Field Howitzer M 39. By changing
the axle the gun could be either horse or motor drawn. Tested in 1938.
Introduced in 1939. Because of its wide range of traverse and its armour
piercing quality it was also used as an anti-tank weapon.

Fieldgun 85mm M 52   Czechoslovakia

Country of origin: Czechoslovakia
Manufacturer: state
Designation: 85mm Gun M 52
Calibre (mm): 85
Length of barrel (x calibre): 56
Depression/Elevation (^{o}): -6 +38
Traverse (^{o}): 60
Weight in firing position (kg): 2,095
Weight of projectile (kg): 9.3
Muzzle velocity (m/sec): 805
Maximum range (km): 16.2
Gun carriage: Split trail
Gun shield: Yes
Mode of traction: Motor drawn
Introduced into: Czechoslovakia

Remarks:

Czechoslovakia's own development. The gun is similar in performance to
the Soviet 85mm Fieldgun M 45. Used mainly as an anti-tank weapon.

Country of origin: Czechoslovakia
Manufacturer: Škoda
Designation: 10cm Light Field Howitzer M 14/19
Calibre (mm): 100
Length of barrel (x calibre): 24
Depression/Elevation (°): -7.5 +50
Traverse (°): 6
Weight in firing position (kg): 1,548
Weight of projectile (kg): 16.0
Muzzle velocity (m/sec): 395
Maximum range (km): 9.8
Gun carriage: Box trail
Gun shield: Yes
Mode of traction: Horse drawn
Introduced into: Czechoslovakia, Hungary

Remarks:

A modification of the 100mm Field Howitzer M 14 (Škoda) from 1921/23. Until 1938 the main gun of the Czechoslovakian divisional artillery.

Country of origin: Czechoslovakia
Manufacturer: Škoda
Designation: 10cm Light Field Howitzer M 28
Calibre (mm): 100
Length of barrel (x calibre): 25
Depression/Elevation (°): -8 +80
Traverse (°): 11
Weight in firing position (kg): 1,660
Weight of projectile (kg): 14.3
Muzzle velocity (m/sec): 449
Maximum range (km): 10.7
Gun carriage: Box trail
Gun shield: Yes
Mode of traction: Horse drawn
Introduced into: Yugoslavia

Remarks:

Standard carriage of the 75mm Fieldgun M 28. Dismantles into three parts
to be tandem drawn for mountain transportation.

Field Howitzer 100mm M 30 (NPH)
Czechoslovakia

Country of origin: Czechoslovakia
Manufacturer: Škoda
Designation: 10cm Light Field Howitzer M 30
Calibre (mm): 100
Length of barrel (x calibre): 25
Depression/Elevation (^{O}): -8 +80
Traverse (^{O}): 8
Weight in firing position (kg): 1,766
Weight of projectile (kg): 16.0
Muzzle velocity (m/sec): 430
Maximum range (km): 10.6
Gun carriage: Box trail
Gun shield: Yes
Mode of traction: Horse or motor drawn
Introduced into: Czechoslovakia

Remarks:

Standard carriage of the 76.5mm Fieldgun M 30. Built in two forms for either horse or motor traction. Dismantles into three parts tandem drawn for mountain transportation. Introduced in 1936. 158 guns were supplied to the units of corps artillery up until 1938.

Field Howitzer 100mm M 39 (H3) Czechoslovakia

Country of origin: Czechoslovakia
Manufacturer: Škoda
Designation: 10cm Light Field Howitzer M 39
Calibre (mm): 100
Length of barrel (x calibre): 29.8
Depression/Elevation (°): -8 +70
Traverse (°): 50
Weight in firing position (kg): 1,960/2,200 *
Weight of projectile (kg): 14.4
Muzzle velocity (m/sec): 525
Maximum range (km): 12.2
Gun carriage: Split trail
Gun shield: Yes
Mode of traction: Horse or motor drawn
Introduced into: Czechoslovakia

Remarks:
Standard carriage with the 76.5mm Fieldgun M 39. By changing the axle
the gun can be either horse or motor drawn. The gun was tested in 1938
and introduced in 1939.

*Horse/motor drawn.

72.

Country of origin: Czechoslovakia
Manufacturer: Škoda
Designation: 7.5cm Mountain Gun M 28
Calibre (mm): 75
Length of barrel (x calibre): 18
Depression/Elevation (^{O}): -9 +50
Traverse (^{O}): 7
Weight in firing position (kg): 710
Weight of projectile (kg): 6.3
Muzzle velocity (m/sec): 425
Maximum range (km): 9.4
Gun carriage: Box trail
Gun shield: Yes
Mode of traction: Horse drawn, by pack animals
Introduced into: Yugoslavia

Remarks:

Dismantles into seven parts. The barrel of the 90mm Mountain Howitzer
M 28 (Škoda) can also be used on the carriage of this gun.

Country of origin: Czechoslovakia
Manufacturer: Škoda
Designation: 7.5cm Mountain Gun M 39
Calibre (mm): 75
Length of barrel (x calibre): 21
Depression/Elevation (⁰): -9.5* +70
Traverse (⁰): 7
Weight in firing position (kg): 820
Weight of projectile (kg): 6.3
Muzzle velocity (m/sec): 480
Maximum range (km): 10.2
Gun carriage: Box trail
Gun shield: Yes
Mode of traction: Horse drawn, by pack animals
Introduced into: Rumania, Iran

Remarks:
Dismantles into eight parts.

*By means of a simple device able to increase to minus 30⁰.

Mountain Howitzer 90mm M 28 (DC)

Country of origin: Czechoslovakia
Manufacturer: Škoda
Designation: 9cm Mountain Howitzer M 28
Calibre (mm): 90
Length of barrel (x calibre): 15.5
Depression/Elevation (^{O}): -9 +50
Traverse (^{O}): 7
Weight in firing position (kg): 710
Weight of projectile (kg): 8.5
Muzzle velocity (m/sec): 350
Maximum range (km): 7.8
Gun carriage: Box trail
Gun shield: Yes
Mode of traction: Horse drawn, pack animals
Introduced into: Yugoslavia

Remarks:

Dismantles into seven parts. The barrel of the 75mm Mountain Gun M 28 (Škoda) can also be used on the carriage of this gun.

Country of origin: Czechoslovakia
Manufacturer: Škoda
Designation: 10cm Mountain Howitzer M 16/19
Calibre (mm): 100
Length of barrel (x calibre): 24
Depression/Elevation (º): -7.5 +70
Traverse (º): 5.5
Weight in firing position (kg): 1,350
Weight of projectile (kg): 16.0
Muzzle velocity (m/sec): 395
Maximum range (km): 9.8
Gun carriage: Box trail
Gun shield: Yes
Mode of traction: Horse drawn, pack animals
Introduced into: Czechoslovakia, Turkey

Remarks:

Three sections tandem drawn. Modification of the 100mm Mountain
Howitzer M 16 (Škoda).

Mountain Howitzer 105mm M 39 (D9)

Country of origin: Czechoslovakia
Manufacturer: Škoda
Designation: 10.5cm Mountain Howitzer M 39
Calibre (mm): 105
Length of barrel (x calibre): 23.9
Depression/Elevation (^{o}): -7.5 +70
Traverse (^{o}): 6
Weight in firing position (kg): 1,400
Weight of projectile (kg): 15.0
Muzzle velocity (m/sec): 450
Maximum range (km): 11.0
Gun carriage: Box trail
Gun shield: Yes
Mode of traction: Horse drawn
Introduced into: Rumania

Remarks:

In three sections tandem drawn. Similar to the 105mm Mountain Howitzer M 39 (D8) built for Afghanistan.

Country of origin: Finland
Manufacturer: Tampella
Designation: 105mm Field Howitzer M/37-10
Calibre (mm): 105
Length of barrel (x calibre): —
Depression/Elevation (^{o}): —
Traverse (^{o}): —
Weight in firing position (kg): 1,700
Weight of projectile (kg): —
Muzzle velocity (m/sec): 550
Maximum range (km): 11.8
Gun carriage: Split trail
Gun shield: Yes
Mode of traction: Motor drawn
Introduced into: Finland

Remarks:
Finnish construction using captured Soviet gun carriages.

Country of origin: Finland
Manufacturer: Tampella
Designation: 105mm Field Howitzer M/61
Calibre (mm): 105
Length of barrel (x calibre): —
Depression/Elevation (°): -6 +45
Traverse (°): 53
Weight in firing position (kg): 1,800
Weight of projectile (kg): 14.9
Muzzle velocity (m/sec): 600
Maximum range (km): 13.4
Gun carriage: Split trail
Gun shield: Yes
Mode of traction: Motor drawn
Introduced into: Finland

Remarks:
Finnish barrels placed in modified gun carriages of captured Soviet Field Howitzers.

Country of origin: France
Manufacturer: Deport
Designation: 7.5cm Fieldgun M 912
Calibre (mm): 75
Length of barrel (x calibre): 28.4
Depression/Elevation (°): -15 +65.5
Traverse (°): 54
Weight in firing position (kg): 1,076
Weight of projectile (kg): 6.5/6.35*
Muzzle velocity (m/sec): 510/500*
Maximum range (km): 7.6/9.0/10.24*
Gun carriage: Split trail
Gun shield: Yes
Mode of traction: Horse and motor drawn
Introduced into: Italy

Remarks:
Made mainly in Italy. At the time of its introduction this gun had a comparatively large angle of elevation and traverse capable of being used for anti-aircraft defence. Ammunition is similar to that of the 75mm Fieldgun M 06 (Krupp).

*With old/new shells.

Country of origin: France
Manufacturer: Schneider
Designation: 75mm Gun M 1897
Calibre (mm): 75
Length of barrel (x calibre): 36
Depression/Elevation (°): -11 +18
Traverse (°): 6
Weight in firing position (kg): 1,160
Weight of projectile (kg): 7.24/7.98 *
Muzzle velocity (m/sec): 529/550 *
Maximum range (km): 8.5/11.2 *
Gun carriage: Box trail
Gun shield: Yes
Mode of traction: Horse or motor drawn
Introduced into: France, numerous other states.

Remarks:

Similar to the 75mm Fieldgun (St Chamond). Designed by Deport (Arsenal Puteaux). The main weapon of the French Artillery in World War I. At the beginning of the war 3,840 were in use by the troops, by the end—5,484. Altogether during the war 17,000 were produced. The first gun with recoil barrel.

* Shell M 1900/M 1917 or M 1918, ALR/2.

This gun is the most widely used of all fieldguns and was still in use during World War II. In this it stands as being the fieldgun in longest use by the troops. In Poland it was known as the 75mm Fieldgun M 97/17.

Partly rebuilt as the 75mm Fieldgun M 97/33 (Schneider), with split trail gun carriage. Later sold to Brazil.

Fieldgun 75mm M 97 built in 1917 with pneumatic tyres.

Country of origin: France
Manufacturer: Schneider
Designation: Gun 75mm M 97/33
Calibre (mm): 75
Length of barrel (x calibre): 36
Depression/Elevation (^{o}): -6 +50
Traverse (^{o}): 58
Weight in firing position (kg): 1,500
Weight of projectile (kg): 6.195
Muzzle velocity (m/sec): 575
Maximum range (km): 11.1
Gun carriage: Split trail
Gun shield: Yes
Mode of traction: Horse and motor drawn
Introduced into: France

Remarks:
Reconstruction of the 75mm Fieldgun M 97 (Schneider) with a new gun
carriage.

Country of origin: France
Manufacturer: Schneider
Designation: 7.5cm Fieldgun M 904
Calibre (mm): 75
Length of barrel (x calibre): 31.4
Depression/Elevation (°): -5 + 16
Traverse (°): 6
Weight in firing position (kg): 1,077
Weight of projectile (kg): 6.5
Muzzle velocity (m/sec): 500
Maximum range (km): 6.0
Gun carriage: Box trail
Gun shield: Yes
Mode of traction: Horse drawn
Introduced into: Portugal

Remarks:
144 of these guns were delivered.

Country of origin: France
Manufacturer: Schneider
Designation: 7.5cm Fieldgun M 05
Calibre (mm): 75
Length of barrel (x calibre): 32
Depression/Elevation (⁰): -5 + 15
Traverse (⁰): 6
Weight in firing position (kg): 1,027
Weight of projectile (kg): 6.50
Muzzle velocity (m/sec): 500
Maximum range (km): 5.9/8.0 *
Gun carriage: Box trail
Gun shield: Yes
Mode of traction: Horse drawn
Introduced into: Bulgaria

Remarks:
324 guns were delivered in 1906/07. Partly replaced in World War I by German fieldguns.

* Original/later increase.

Country of origin: France
Manufacturer: Schneider
Designation: 7.5cm Fieldgun M 06
Calibre (mm): 75
Length of barrel (x calibre): 30
Depression/Elevation (º): -3 + 16
Traverse (º): 6
Weight in firing position (kg): 1,040
Weight of projectile (kg): 6.5
Muzzle velocity (m/sec): 500
Maximum range (km): 5.8/8.5*
Gun carriage: Box trail
Gun shield: Yes
Mode of traction: Horse drawn
Introduced into: Spain, Greece

Remarks:

200 guns were supplied to Spain. These equipped the gun regiment of the divisional artillery until the Civil war. During the Civil War this gun was replaced by numerous other weapons.
144 guns were supplied to Greece.

*With old/new shells.

Country of origin: France
Manufacturer: Schneider
Designation: 7.5cm Fieldgun M 7
Calibre (mm): 75
Length of barrel (x calibre): 31.4
Depression/Elevation (^{o}): -8 +16
Traverse (^{o}): 6
Weight in firing position (kg): 1,083
Weight of projectile (kg): 6.5
Muzzle velocity (m/sec): 500
Maximum range (km): 7.3
Gun carriage: Box trail
Gun shield: Yes
Mode of traction: Horse drawn
Introduced into: Serbia

Remarks:
Similar to M 7A. 188 guns were ordered in 1906. Replacement by the
75mm Fieldgun M 28 (Škoda) was ordered, but only partly carried out.

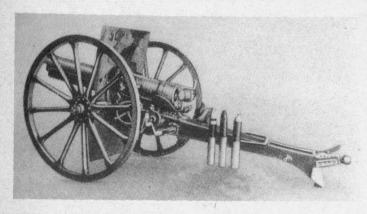

Country of origin: France
Manufacturer: Schneider
Designation: 75mm Gun M 1912
Calibre (mm): 75
Length of barrel (x calibre): 25.4
Depression/Elevation (^{o}): -9 +17
Traverse (^{o}): 6
Weight in firing position (kg): 965
Weight of projectile (kg): 7.24
Muzzle velocity (m/sec): 500
Maximum range (km): 7.5
Gun carriage: Box trail
Gun shield: Yes
Mode of traction: Horse drawn
Introduced into: France, Serbia

Remarks:

Similar to the 75mm Fieldgun M 14 (Schneider), barrel length L/31.
Introduced into France for the mounted batteries. The gun did not come
up to expectations and was replaced during World War I by the 75mm
Fieldgun M 97 (Schneider).

Country of origin: France
Manufacturer: Schneider
Designation: 7.5cm Fieldgun M 22
Calibre (mm): 75
Length of barrel (x calibre): 31.3
Depression/Elevation (°): -5 +43
Traverse (°): 50
Weight in firing position (kg): 1,320
Weight of projectile (kg): 6.3
Muzzle velocity (m/sec): 600
Maximum range (km): 12.0
Gun carriage: Split trail
Gun shield: Yes
Mode of traction: Horse drawn
Introduced into: Finland

Remarks:
One of the first fieldguns of the new construction to be introduced after
World War I.

Country of origin: France
Manufacturer: de Bange
Designation: Gun 90mm de Bange
Calibre (mm): 90
Length of barrel (x calibre): 25.3
Depression/Elevation (º): -5 +26
Traverse (º): 0
Weight in firing position (kg): 1,220/1,200 *
Weight of projectile (kg): 8.0/8.3 *
Muzzle velocity (m/sec): 455/500 *
Maximum range (km): 7.0/9.7 *
Gun carriage: Box trail
Gun shield: No
Mode of traction: Horse drawn
Introduced into: France, Serbia

Remarks:

Recoil carriage gun. At the beginning of World War I this gun was used in France to supplement the 75mm Fieldgun M 97 (Schneider), equipment stored for 100 batteries were again put in service. The range was increased by enlarging the projectile charge and by using a new form of shell.

* Before/after modification.

90

Country of origin: France
Manufacturer: Schneider
Designation: 8.5cm Gun
Calibre (mm): 85
Length of barrel (x calibre): 34.8
Depression/Elevation ($^{\circ}$): -6 +65
Traverse ($^{\circ}$): 54
Weight in firing position (kg): 1,970
Weight of projectile (kg): 10.0
Muzzle velocity (m/sec): 675
Maximum range (km): 15.0
Gun carriage: Split trail
Gun shield: Yes
Mode of traction: Motor drawn
Introduced into: Greece

Remarks:
Used in the regiments of the heavy artillery.

Country of origin: France
Manufacturer: Atelier Bourges
Designation: 10.5cm Field Howitzer 1935 B
Calibre (mm): 105
Length of barrel (x calibre): 16.7
Depression/Elevation (^{O}): -6 +50
Traverse (^{O}): 58
Weight in firing position (kg): 1,627
Weight of projectile (kg): 15.67
Muzzle velocity (m/sec): 442
Maximum range (km): 10.3
Gun carriage: Split trail
Gun shield: Yes
Mode of traction: Horse and motor drawn
Introduced into: France

Remarks:

410 of these guns became available in 1939. At the beginning of World
War II one battalion of the artillery regiment of the division was equipped
with such guns. This marked France's abandonment of the exclusive
equipping of the field artillery with the 75mm Fieldgun M 97 (Schneider).

Country of origin: France
Manufacturer: Schneider
Designation: 10.5cm Field Howitzer M 10
Calibre (mm): 105
Length of barrel (x calibre): 14
Depression/Elevation (^{O}): -3 +43
Traverse (^{O}): 6
Weight in firing position (kg): 1,150
Weight of projectile (kg): 16.4
Muzzle velocity (m/sec): 300
Maximum range(km): 6.2
Gun carriage: Box trail
Gun shield: Yes
Mode of traction: Horse drawn
Introduced into: Rumania

Remarks:

Country of origin: France
Manufacturer: Schneider
Designation: 10.5cm Field Howitzer 1934 S
Calibre (mm): 105
Length of barrel (x calibre): 20
Depression/Elevation (^{O}): -8 +43
Traverse (^{O}): 42
Weight in firing position (kg): 1,722
Weight of projectile (kg): 15.85
Muzzle velocity (m/sec): 465
Maximum range (km): 10.7
Gun carriage: Split trail
Gun shield: Yes
Mode of traction: Horse and motor drawn
Introduced into: France

Remarks:

Field Howitzer 120mm M 90   France

Country of origin: France
Manufacturer: Baquet
Designation: Short Gun 120mm M 90
Calibre (mm): 120
Length of barrel (x calibre): 14.2
Depression/Elevation (°): -12 +44
Traverse (°): 10
Weight in firing position (kg): 1,475
Weight of projectile (kg): 18.0/20.0
Muzzle velocity (m/sec): 290/284
Maximum range (km): 5.8/5.6
Gun carriage: Box trail
Gun shield: No
Mode of traction: Horse drawn
Introduced into: France, Belgium

Remarks:
Withdrawn shortly after the outbreak of World War I.

Country of origin: France
Manufacturer: St Chamond
Designation: 12cm Quickfiring Howitzer
Calibre (mm): 120
Length of barrel (x calibre): —
Depression/Elevation (°): -10 +40
Traverse (°): —
Weight in firing position (kg): 1,300
Weight of projectile (kg): 20.0
Muzzle velocity (m/sec): 300
Maximum range (km): —
Gun carriage: Box trail
Gun shield: Yes
Mode of traction: Horse drawn
Introduced into: Belgium

Remarks:
Introduced at the beginning of World War I.

Field Howitzer 120mm M 97   France

Country of origin: France
Manufacturer: Schneider
Designation: Field Howitzer 120mm M 97
Calibre (mm): 120
Length of barrel (x calibre): 12
Depression/Elevation (°): -5 +45
Traverse (°): 4
Weight in firing position (kg): 1,140
Weight of projectile (kg): 21.0
Muzzle velocity (m/sec): 300
Maximum range (km): 6.65
Gun carriage: Box trail
Gun shield: No
Mode of traction: Horse drawn
Introduced into: Serbia

Remarks:
At the end of the 19th Century Serbia bought guns for six batteries.

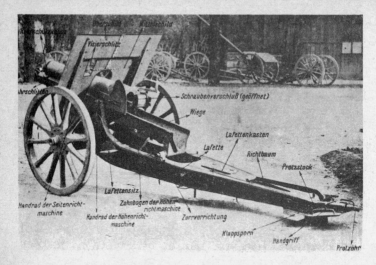

Country of origin: France
Manufacturer: Schneider
Designation: 12.2cm Howitzer M 10 (48 line Howitzer)
Calibre (mm): 121.9
Length of barrel (x calibre): 12.8
Depression/Elevation (^{0}): -3 +45
Traverse (^{0}): 9
Weight in firing position (kg): 1,324
Weight of projectile (kg): 23.0
Muzzle velocity (m/sec): 335
Maximum range (km): 7.68
Gun carriage: Box trail
Gun shield: Yes
Mode of traction: Horse drawn
Introduced into: Russia, Finland

Remarks:
Together with the 122mm Field Howitzer M 09 (Krupp), this was the most
important field howitzer at the beginning of World War I. Modified in the
Soviet Union, it became the 122mm Field Howitzer M 10/30 (state).

Country of origin: France
Manufacturer: Schneider
Designation: 12cm Light Field Howitzer M 11
Calibre (mm): 120
Length of barrel (x calibre): 13
Depression/Elevation (^{O}): -3 +43
Traverse (^{O}): 5
Weight in firing position (kg): 1,385
Weight of projectile (kg): 21.0
Muzzle velocity (m/sec): 330
Maximum range (km): 6.7/8.0 *
Gun carriage: Box trail
Gun shield: Yes
Mode of traction: Horse drawn
Introduced into: Bulgaria, Serbia, Belgium

Remarks:

Similar to the M 7, 36 of which had been ordered by Bulgaria.

* Original/later increase.

Country of origin: France
Manufacturer: Schneider
Designation: Howitzer 120mm TR
Calibre (mm): 120
Length of barrel (x calibre): 13
Depression/Elevation (^{o}): -3 +43
Traverse (^{o}): 5
Weight in firing position (kg): 1,476
Weight of projectile (kg): 18.7—19.3
Muzzle velocity (m/sec): 350
Maximum range (km): 8.3
Gun carriage: Box trail
Gun shield: Yes
Mode of traction: Horse drawn
Introduced into: France

Remarks:

Introduced in 1915, but only a small number reached the front. Production was stopped as it was considered in France that a light field howitzer was no longer necessary.

Country of origin: France
Manufacturer: Schneider-Ducrest
Designation: Mountain Gun 65mm M 06
Calibre (mm): 65
Length of barrel (x calibre): 20.2
Depression/Elevation (°): -9.5, +35
Traverse (°): 6
Weight in firing position (kg): 400
Weight of projectile (kg): 3.81—4.45
Muzzle velocity (m/sec): 330
Maximum range (km): 5.5
Gun carriage: Box trail
Gun shield: No
Mode of traction: Horse drawn, by pack animals
Introduced into: France, Poland

Remarks:

Dismantles into four parts. At the beginning of World War I there were 120
of these counter-recoil guns in the French units, by the end of the war the
number was 96. After the war it was used by the French as an Infantry gun
in the Colonies. The only counter-recoil gun to be introduced in quantity.

Country of origin: France
Manufacturer: Schneider
Designation: 7cm Mountain Gun M 1907
Calibre (mm): 70
Length of barrel (x calibre): 17.1
Depression/Elevation (°): -10 +20
Traverse (°): 4.5
Weight in firing position (kg): 436
Weight of projectile (kg): 5.1
Muzzle velocity (m/sec): 330
Maximum range (km): 5.5
Gun carriage: Box trail
Gun shield: Yes
Mode of traction: Horse drawn, by pack animals
Introduced into: Portugal

Remarks:
Dismantles into five parts.

Country of origin: France
Manufacturer: Schneider
Designation: 7cm Mountain Gun M 07
Calibre (mm): 70
Length of barrel (x calibre): 17.14
Depression/Elevation (^{O}): -10 +30
Traverse (^{O}): 5
Weight in firing position (kg): 495
Weight of projectile (kg): 5.0
Muzzle velocity (m/sec): 300
Maximum range (km): 5.0
Gun carriage: Box trail
Gun shield: Yes
Mode of traction: Horse drawn, by pack animals
Introduced into: Serbia

Remarks:
Dismantles into five parts. 36 guns were delivered to Serbia.

Country of origin: France
Manufacturer: Schneider
Designation ; 7cm Mountain Gun M 08
Calibre (mm): 70
Length of barrel (x calibre): 17.1
Depression/Elevation (^{O}): -10 +20
Traverse (^{O}): 4.5
Weight in firing position (kg): 508
Weight of projectile (kg): 5.3
Muzzle velocity (m/sec): 300
Maximum range (km): 5.0
Gun carriage: Box trail
Gun shield: Yes
Mode of traction: Horse drawn, by pack animals
Introduced into: Spain, Italy

Remarks:

Dismantles into five parts. 48 guns delivered to Spain, who used them as Infantry guns from 1923 on. Italy received these guns during World War I and later also used them as Infantry guns.

Country of origin: France
Manufacturer: Schneider
Designation: 7.5cm Mountain Gun M 07
Calibre (mm): 75
Length of barrel (x calibre): 16
Depression/Elevation (^{O}): -10 +30
Traverse (^{O}): 4.5
Weight in firing position (kg): 506
Weight of projectile (kg): 5.1
Muzzle velocity (m/sec): 330
Maximum range (km): 6.0
Gun carriage: Box trail
Gun shield: Yes
Mode of traction: Horse drawn, by pack animals
Introduced into: Bulgaria

Remarks:
Dismantles into five parts. 36 guns were delivered to Bulgaria.

Country of origin: France
Manufacturer: Schneider
Designation: 7.5cm Mountain Gun
Calibre (mm): 75
Length of barrel (x calibre): 16
Depression/Elevation (°): -10 +20
Traverse (°): 4.5
Weight in firing position (kg): 513
Weight of projectile (kg): 5.3
Muzzle velocity (m/sec): 300
Maximum range (km): 6.0
Gun carriage: Box trail
Gun shield: Yes
Mode of traction: Horse drawn, by pack animals
Introduced into: Turkey

Remarks:
Dismantles into five parts.

Country of origin: France
Manufacturer: Schneider/Danglis
Designation: 7.5cm Mountain Gun M 06/09
Calibre (mm): 75
Length of barrel (x calibre): 16.7
Depression/Elevation (^{o}): -7.4 +20
Traverse (^{o}): 4.3
Weight in firing position (kg): 600
Weight of projectile (kg): 6.5
Muzzle velocity (m/sec): 350
Maximum range (km): 6.5
Gun carriage: Box trail
Gun shield: No
Mode of traction: Horse drawn, by pack animals
Introduced into: Greece

Remarks:

Dismantles into five parts. Designed by the Greek Colonel Danglis and
built by Schneider.

Country of origin: France
Manufacturer: Schneider
Designation: 7.6cm Mountain Gun M 09
Calibre (mm): 76.2
Length of barrel (x calibre): 16.5
Depression/Elevation (^{O}): -10 +35 (+29.5*)
Traverse (^{O}): 4.5
Weight in firing position (kg): 626
Weight of projectile (kg): 6.23
Muzzle velocity (m/sec): 387
Maximum range (km): 8.55
Gun carriage: Box trail
Gun shield: Yes
Mode of traction: Horse drawn, by pack animals
Introduced into: Russia, Finland

Remarks:
Dismantles into six parts. The gun was originally built to have a muzzle
velocity of 350m/sec. This was increased by Russia to 387m/sec.

In Russia.

Country of origin: France
Manufacturer: Schneider
Designation: 75mm Mountain Gun M 19
Calibre (mm): 75
Length of barrel (x calibre): 18.6
Depression/Elevation (^{o}): -10 +40
Traverse (^{o}): 10
Weight in firing position (kg): 659
Weight of projectile (kg): 6.33
Muzzle velocity (m/sec): 440
Maximum range (km): 9.5
Gun carriage: Box trail
Gun shield: Yes
Mode of traction: Horse drawn, by pack animals
Introduced into: France, Greece, Poland

Remarks:

Dismantles into seven parts.

Mountain Gun 75mm M 28

Country of origin: France
Manufacturer: Schneider
Designation: 75mm Mountain Gun M 28
Calibre (mm): 75
Length of barrel (x calibre): 18.6
Depression/Elevation (°): -10 +40
Traverse (°): 10
Weight in firing position (kg): 660
Weight of projectile (kg): 6.33
Muzzle velocity (m/sec): 440
Maximum range (km): 9.5
Gun carriage: Box trail
Gun shield: Yes
Mode of traction: Horse drawn, by pack animals
Introduced into: France

Remarks:
Dismantles into seven parts.

Country of origin: France
Manufacturer: Schneider
Designation: 10.5cm Mountain Howitzer M 09
Calibre (mm): 105
Length of barrel (x calibre): 10.5
Depression/Elevation (^{O}): -0 +60
Traverse (^{O}): 5
Weight in firing position (kg): 730
Weight of projectile (kg): 12.0
Muzzle velocity (m/sec): 300
Maximum range (km): 6.0
Gun carriage: Box trail
Gun shield: Yes
Mode of traction: Horse drawn, by pack animals
Introduced into: Russia

Remarks:

Dismantles into seven parts. In use in the Soviet Union until World War II.

Country of origin: France
Manufacturer: Schneider
Designation: 105mm Mountain Howitzer M 19
Calibre (mm): 105
Length of barrel (x calibre): 12.4
Depression/Elevation (°): -0 +43
Traverse (°): 9
Weight in firing position (kg): 750
Weight of projectile (kg): 12.4
Muzzle velocity (m/sec): 350
Maximum range (km): 7.85
Gun carriage: Box trail
Gun shield: Yes
Mode of traction: Horse drawn, by pack animals
Introduced into: France, Spain

Remarks:

Dismantles into eight parts. Mountain Howitzer with barrel which can be dismantled.

Country of origin: France
Manufacturer: Schneider
Designation: 105mm Mountain Howitzer M 28
Calibre (mm): 105
Length of barrel (x calibre): 12.4
Depression/Elevation (^{0}): -0 +40
Traverse (^{0}): 9
Weight in firing position (kg): 745
Weight of projectile (kg): 12.0
Muzzle velocity (m/sec): 330
Maximum range (km): 7.3
Gun carriage: Box trail
Gun shield: Yes
Mode of traction: Horse drawn, by pack animals
Introduced into: France

Remarks:

Dismantles into seven parts.

Country of origin: Germany
Manufacturer: Ehrhardt
Designation: 7.62cm (15 Pdr) Fieldgun
Calibre (mm): 76.2
Length of barrel (x calibre): 30
Depression/Elevation (^{o}): -10 + 17
Traverse (^{o}): 7
Weight in firing position (kg): 980
Weight of projectile (kg): 6.58
Muzzle velocity (m/sec): 500
Maximum range (km): 5.5
Gun carriage: Pole trail
Gun shield: No
Mode of traction: Horse drawn
Introduced into: Great Britain

Remarks:
108 of these guns were bought during the Boer War. Replaced by the
83.8mm Fieldgun M 03 (Vickers).

Country of origin: Germany
Manufacturer: Ehrhardt
Designation: 7.5cm Fieldgun M 01
Calibre (mm): 75
Length of barrel (x calibre): 31
Depression/Elevation (^{O}): -7 + 15.5
Traverse (^{O}): 7
Weight in firing position (kg): 1,023
Weight of projectile (kg): 6.5
Muzzle velocity (m/sec): 500
Maximum range (km): 6.0 *
Gun carriage: Pole trail
Gun shield: Yes
Mode of traction: Horse drawn
Introduced into: Norway

Remarks:
132 were bought. In use until the beginning of World War II.

*Later, with an entrenched trail spade, range of 10.6km.

Country of origin: Germany
Manufacturer: Krupp
Designation: 7.5cm Fieldgun M 02
Calibre (mm): 75
Length of barrel (x calibre): 30
Depression/Elevation (^{O}): -9 +15
Traverse (^{O}): 7
Weight in firing position (kg): 1,079
Weight of projectile (kg): 6.8/8.671*
Muzzle velocity (m/sec): 510/546*
Maximum range (km): 6.0/10.0 *
Gun carriage: Box trail
Gun shield: Yes
Mode of traction: Horse drawn
Introduced into: Denmark

Remarks:

* With old/new ammunition and entrenched trail spade.

Country of origin: Germany
Manufacturer: Krupp
Designation: 7.5cm Fieldgun M 02
Calibre (mm): 75
Length of barrel (x calibre): 30
Depression/Elevation (⁰): -8 + 16
Traverse (⁰): 6
Weight in firing position (kg): 975
Weight of projectile (kg): 6.5
Muzzle velocity (m/sec): 500
Maximum range (km): 7.0
Gun carriage: Box trail
Gun shield: Yes
Mode of traction: Horse drawn
Introduced into: Sweden

Remarks:
Partly built under licence in Sweden. Later rebuilt, see Sweden. About two-thirds of the guns were produced in Sweden. Later rebuilt by Bofors as the 75mm Fieldgun M 02/33.

Country of origin: Germany
Manufacturer: Krupp
Designation: 7.5cm Fieldgun
Calibre (mm): 75
Length of barrel (x calibre): 30
Depression/Elevation (^{O}): -8 +16
Traverse (^{O}): 4
Weight in firing position (kg): 995
Weight of projectile (kg): 6.35
Muzzle velocity (m/sec): 485
Maximum range (km): 6.5
Gun carriage: Box trail
Gun shield: Yes
Mode of traction: Horse drawn
Introduced into: Switzerland

Remarks:

Later rebuilt, see Switzerland. Rebuilt by Sulzer after World War I. In use in this form until the end of World War II.

Country of origin: Germany
Manufacturer: Krupp
Designation: 7.5cm Fieldgun M 03
Calibre (mm): 75
Length of barrel (x calibre): 30
Depression/Elevation (°): -8 + 16
Traverse (°): 4
Weight in firing position (kg): 1,070
Weight of projectile (kg): 6.5
Muzzle velocity (m/sec): 500
Maximum range (km): 8.0
Gun carriage: Box trail
Gun shield: Yes
Mode of traction: Horse drawn
Introduced into: Rumania

Remarks:

Up to 1908 360 of these guns were delivered. They were still in use in World War II. The 75mm Fieldgun M 28 (Škoda) was intended to replace this gun.

Country of origin: Germany
Manufacturer: Krupp
Designation: 7.5cm Fieldgun M 03
Calibre (mm): 75
Length of barrel (x calibre): 30
Depression/Elevation (^{0}): -7 +16.5
Traverse (^{0}): 8
Weight in firing position (kg): 990
Weight of projectile (kg): 6.0
Muzzle velocity (m/sec): 500
Maximum range (km): 6.4
Gun carriage: Box trail
Gun shield: Yes
Mode of traction: Horse drawn
Introduced into: Netherlands

Remarks:
Later rebuilt, see Netherlands. 204 guns were delivered. Reconstructed in the Netherlands after World War I.

Country of origin: Germany
Manufacturer: Krupp
Designation: 7.5cm Fieldgun M 03
Calibre (mm): 75
Length of barrel (x calibre): 30
Depession/Elevation (°): -5 + 15
Traverse (°): 7
Weight in firing position (kg): 1,000
Weight of projectile (kg): 6.35
Muzzle velocity (m/sec): 500
Maximum range (km): 5.9/8.0 *
Gun carriage: Box trail
Gun shield: Yes
Mode of traction: Horse drawn
Introduced into: Turkey

Remarks:

558 guns delivered between 1903 and 1907. Improved in 1910.

* Before/after improvement.

Country of origin: Germany
Manufacturer: Krupp/Ehrhardt
Designation: 7.7cm Fieldgun 96 (New Type)
Calibre (mm): 77
Length of barrel (x calibre): 27.3
Depression/Elevation (^{o}): -13 + 15
Traverse (^{o}): 8
Weight in firing position (kg): 925 *
Weight of projectile (kg): 6.85
Muzzle velocity (m/sec): 465
Maximum range (km): 7.8
Gun carriage: Box trail
Gun shield: Yes
Mode of traction: Horse drawn
Introduced into: Germany, Bulgaria, Turkey

Remarks:

Principal weapon of the divisional artillery at the beginning of World War I.
5,068 were in use by the troops, and by the end of the war, 3,744 were still
in service. Originally developed in 1904 from the 96 Fieldgun which had a
rigid gun carriage system, the recoil being checked by a rope brake and a
hinged trail spade.

* 875 kg for the horse artillery.

Country of origin: Germany
Manufacturer: Krupp
Designation: 7.5cm Fieldgun TR
Calibre (mm): 75
Length of barrel (x calibre): 30
Depression/Elevation (^{o}): -10 +16
Traverse (^{o}): 6
Weight in firing position (kg): 1,035
Weight of projectile (kg): 6.5
Muzzle velocity (m/sec): 500
Maximum range: 8.0
Gun carriage: Box trail
Gun shield: Yes
Mode of traction: Horse drawn
Introduced into: Belgium

Remarks:

Partly built in Belgium by FRC and Cockerill. The gun continued to be built
in Belgium up until World War I, the materials for the barrel being supplied
by Great Britain.

Country of origin: Germany
Manufacturer: Krupp
Designation: 7.5cm Fieldgun Meiji 38
Calibre (mm): 75
Length of barrel (x calibre): 30
Depression/Elevation (^{o}): -8 +16.5
Traverse (^{o}): 7
Weight in firing position (kg): 950
Weight of projectile (kg): 6.5
Muzzle velocity (m/sec): 500
Maximum range (km): 7.9
Gun carriage: Box trail
Gun shield: Yes
Mode of traction: Horse drawn
Introduced into: Japan

Remarks:

Partly built in Japan under licence. Introduced after the Russian/Japanese war as a replacement for the 75mm Fieldgun M 98 (Meiji 31), with recoil carriage. At the beginning of 1905 about 400 of these guns were delivered by Krupp. Later a further 300 were built in Japan by Osaka Arsenal from material supplied by Krupp. It was the main weapon of the divisional artillery until the beginning of the thirties.

Fieldgun 75mm M 06

Country of origin: Germany
Manufacturer: Krupp
Designation: 7.5cm Fieldgun M 06
Calibre (mm): 75
Length of barrel (x calibre): 30
Depression/Elevation (^{o}): -10 + 16
Traverse (^{o}): 7
Weight in firing position (kg): 1,032/1,015 — 1,080 *
Weight of projectile (kg): 6.5/6.35 †
Muzzle velocity (m/sec): 510/500 †
Maximum range (km): 6.8/10.25 †
Gun carriage: Box trail
Gun shield: Yes
Mode of traction: Horse or motor drawn
Introduced into: Italy

Remarks:

Partly completed in Italy. The first 39 batteries were delivered by Krupp, more were built in Italy from already produced components. In the Italian army the gun replaced the old 75mm A Fieldgun with recoil carriage.

*Originally/later horse or motor drawn.
†With old/new shells.

Fieldgun 75mm M 11*

Country of origin: Germany
Manufacturer: Krupp
Designation: 7.5cm Fieldgun M 911*
Calibre(mm): 75
Length of barrel (x calibre): 30
Depression/Elevation (°): -12 +18.5
Traverse (°): 7
Weight in firing position (kg): 958/943†
Weight of projectile (kg): 6.5/6.35 ‡
Muzzle velocity (m/sec): 510/500 ‡
Maximum range (km): 7.6/10.24 ‡
Gun carriage: Box trail
Gun shield: Yes
Mode of traction: Horse drawn
Introduced into: Italy

Remarks:

Only 12 batteries were ordered. Primarily intended for use by the horse artillery, but also partly used by the mobile units and later by the 'fast divisions'.

* Also designated M 06/12
† When used by the mobile/horse batteries.
‡ With old/new shells.

Country of origin: Germany
Manuacturer: Krupp
Designation: 7.7cm Fieldgun 96/16
Calibre (mm): 77
Length of barrel (x calibre): 27
Depression/Elevation (°): -12 +15
Traverse (°): 8
Weight in firing position (kg): 1,020
Weight of projectile (kg): 6.25/6.85
Muzzle velocity (m/sec): 477/465
Maximum range (km): 8.025/8.4
Gun carriage: Box trail
Gun shield: Yes
Mode of traction: Horse drawn
Introduced into: Germany

Remarks:
The gun differs only slightly from the 77mm Fieldgun M 96, New Type
(Krupp/Ehrhardt).

Country of origin: Germany
Manufacturer: Krupp
Designation: 7.5cm Fieldgun 16 (New Type)
Calibre (mm): 75
Length of barrel (x calibre): 36
Depression/Elevation (^{o}): -9 +40
Traverse (^{o}): 4
Weight in firing position (kg): 1,460
Weight of projectile (kg): 6.37
Muzzle velocity (m/sec): 660
Maximum range (km): 12.3
Gun carriage: Box trail
Gun shield: Yes
Mode of traction: Horse drawn
Introduced into: Germany

Remarks:

Introduced into the horse artillery in 1934 as a replacement for the 77mm
Fieldgun M 16 (Rheinmetall).

Country of origin: Germany
Manufacturer: Krupp
Designation: 7.5cm light Fieldgun 18
Calibre (mm): 75
Length of barrel (x calibre): 26
Depression/Elevation (°): -5 +45
Traverse (°): 60
Weight in firing position (kg): 1,120
Weight of projectile (kg): —
Muzzle velocity (m/sec): 475
Maximum range (km): 9.425
Gun carriage: Split trail
Gun shield: Yes
Mode of traction: Horse drawn
Introduced into: Germany

Remarks:

Developed in 1930/31. First introduced into the horse artillery regiment of
the cavalry brigade in 1938 as a replacement for the 75mm Fieldgun M 16,
New Type (Krupp).

Country of origin: Germany
Manufacturer: Krupp
Designation: 7.5cm Fieldgun 38
Calibre (mm): 75
Length of barrel (x calibre): 34
Depression/Elevation (°): -5 +45
Traverse (°): 55
Weight in firing position (kg): 1,380
Weight of projectile (kg): —
Muzzle velocity (m/sec): 580
Maximum range (km): 11.3
Gun carriage: Split trail
Gun shield: Yes
Mode of traction: Horse and motor drawn
Introduced into: Germany, Brazil

Remarks:

Developed for an order from the Brazilian Government. About 80 of the
existing guns by Krupp were taken by the German army to replace the
75mm Fieldgun M 18 (Krupp). At the end of the war they were still in use
by the 'Volksgrenadier' Divisions.

Country of origin: Germany
Manufacturer: Rheinmetall
Designation: 7.7cm Fieldgun 16
Calibre (mm): 77
Length of barrel (x calibre): 35
Depression/Elevation (º): -10 + 40
Traverse (º): 4
Weight in firing position (kg): 1,325
Weight of projectile (kg): 7.2/5.89*
Muzzle velocity (m/sec): 545/602*
Maximum range (km): 9.1/10.7*
Gun carriage: Box trail
Gun shield: Yes
Mode of traction: Horse drawn
Introduced into: Germany, Bulgaria, Finland, Turkey

Remarks:

Partly replaced the 77mm Fieldgun M 96, New Type (Krupp/Ehrhardt). At the end of World War I it was found that 3,020 of these guns were still in use by the Germany artillery units. The construction is similar to that of its predecessors.

*With shell/C—projectile from 1917.

Country of origin: Germany
Manufacturer: Rheinmetall
Designation: 7.5cm Fieldgun 7 M 85
Calibre (mm): 75
Length of barrel (x calibre): 42.5
Depression/Elevation ($^{\circ}$): -5 +42
Traverse ($^{\circ}$): 30.5
Weight in firing position (kg): 1,778
Weight of projectile (kg): 5.74
Muzzle velocity (m/sec): 550
Maximum range (km): 10.275
Gun carriage: Split trail (tubular)
Gun shield: Yes
Mode of traction: Motor drawn
Introduced into: Germany

Remarks:

An attempt to combine the fieldgun and anti-tank weapon. The barrel and cradle of the 75mm Anti-Tank gun M 40 (Rheinmetall), in the gun carriage of the 105mm Field Howitzer M 18/40 (Rheinmetall).

Country of origin: Germany
Manufacturer: Krupp
Designation: 10.5cm Light Field Howitzer 98/09
Calibre (mm): 105
Length of barrel (x calibre): 16
Depression/Elevation (°): -10 +40
Traverse (°): 4
Weight in firing position (kg): 1,225
Weight of projectile (kg): 15.7
Muzzle velocity (m/sec): 302
Maximum range (km): 6.3
Gun carriage: Box trail
Gun shield: Yes
Mode of traction: Horse drawn
Introduced into: Germany, Bulgaria, Turkey

Remarks:

Arose out of the rebuilding of the Light Field Howitzer 98, which was
again a gun with recoil carriage. Introduced at the end of 1909. At the
beginning of World War I 1,260 were to be found in the German artillery
units, by the end of the war the number was 1,144.

Country of origin: Germany
Manufacturer: Krupp
Designation: 10.5cm Howitzer M 12/16
Calibre (mm): 105
Length of barrel (x calibre): 14
Depression/Elevation (^{O}): -5 +43
Traverse (^{O}): 6
Weight in firing position (kg): 1,155
Weight of projectile (kg): 14.0
Muzzle velocity (m/sec): 300
Maximum range (km): 6.5
Gun carriage: Box trail
Gun shield: Yes
Mode of traction: Horse drawn
Introduced into: Rumania, Bulgaria

Remarks:

60 guns ordered by Rumania.

Country of origin: Germany
Manufacturer: Krupp
Designation: 10.5cm Light Field Howitzer Krupp
Calibre (mm): 104.9
Length of barrel (x calibre): 20
Depression/Elevation (^{O}): -4 +43
Traverse (^{O}): 4
Weight in firing position (kg): 1,500
Weight of projectile (kg): 15.7
Muzzle velocity (m/sec): 430/453 *
Maximum range (km): 8.95/10.2 *
Gun carriage: Box trail
Gun shield: Yes
Mode of traction: Horse drawn
Introduced into: Germany

Remarks:

Introduced in addition to the 105mm Field Howitzer M 16 (Rheinmetall).
At the end of World War I 264 of these guns were in use in the forces. Use
was made of the gun carriage of the 120mm Field Howitzer M 12 (Krupp)
which was produced for the Swiss.

*With FH shell 98/C projectile.

Field Howitzer 105mm M 16

Country of origin: Germany
Manufacturer: Rheinmetall
Designation: 10.5cm Light Field Howitzer 16
Calibre (mm): 104.9
Length of barrel (x calibre): 22
Depression/Elevation (^{o}): -10 +40
Traverse (^{o}): 4
Weight in firing position (kg): 1,380/1,525 *
Weight of projectile (kg): 15.7
Muzzle velocity (m/sec): 400/427/395†
Maximum range (km): 8.4/9.7/9.275 †
Gun carriage: Box trail
Gun shield: Yes
Mode of traction: Horse drawn
Introduced into: Germany, Belgium, Bulgaria, Lithuania, Poland, Turkey

Remarks:

Replaced the 105mm Field Howitzer M 98/09 (Krupp). at the end of World War I 3,004 of these guns were still in use in the German artillery units. The standard weapon of the divisional artillery, until the introduction of the 105mm Field Howitzer M 18 (Rheinmetall). Later found only in isolated instances until 1945.

*After later modifications heavier.
†With FH shells 98/C projectile from 1917/FH shell of the German Army from 1934.

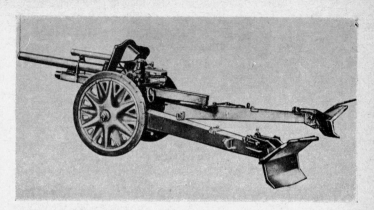

Country of origin: Germany
Manufacturer: Rheinmetall
Designation: 10.5cm Light Field Howitzer 18
Calibre (mm): 104.9
Length of barrel (x calibre): 28
Depression/Elevation (^{o}): -5 +42
Traverse (^{o}): 56
Weight in firing position (kg): 1,915/1,985*
Weight of projectile (kg): 14.8
Muzzle velocity (m/sec): 470
Maximum range (km): 10.675
Gun carriage: Split trail
Gun shield: Yes
Mode of traction: Horse and motor drawn
Introduced into: Germany, Finland, Norway, Sweden, Spain, Hungary

Remarks:

Developed 1928/29. Introduced 1935. Until the start of World War II 4,845
guns were built. They were the standard weapon of the divisional artillery.
It was only partly replaced by the 105mm Field Howitzer M 18M
(Rheinmetall/Krupp) and the 105mm Field Howitzer M 18/40
(Rheinmetall).

* When horse drawn/motor drawn.

Country of origin: Germany
Manufacturer: Rheinmetall/Krupp
Designation: 10.5cm Light Field Howitzer 18M
Calibre (mm): 104.9
Length of barrel (x calibre): 28
Depression/Elevation (°): -5 +42
Traverse (°): 56
Weight in firing position (kg): 2,040
Weight of projectile (kg): 14.8
Muzzle velocity (m/sec): 540
Maximum range (km): 12.325
Gun carriage: Split trail
Gun shield: Yes
Mode of traction: Horse or motor drawn
Introduced into: Germany, many countries during and after World War II.

Remarks:

Similar to the Field Howitzers 105mm, M 18/39 and M 39. Similar in construction to the 105mm Field Howitzer M 18 (Rheinmetall). The 105mm Field Howitzer M 39 was delivered to the Netherlands just before the beginning of the war. The barrel of the Field Howitzer M 18M (Rheinmetall/Krupp) was placed in existing gun carriages, and these guns were known as the 105mm Field Howitzer M 18/39 in the German forces.

Field Howitzer 105mm M 18/40

Germany

Country of origin: Germany
Manufacturer: Rheinmetall
Designation: 10.5cm Light Field Howitzer 18/40
Calibre (mm): 104.9
Length of barrel (x calibre): 28
Depression/Elevation (°): -5 +45
Traverse (°): 60
Weight in firing position (kg): 1,800
Weight of projectile (kg): 14.8
Muzzle velocity (m/sec): 540
Maximum range (km): 12,325
Gun carriage: Split trail (tubular)
Gun shield: Yes
Mode of traction: Horse and motor drawn
Introduced into: Germany, Austria

Remarks:

Constructed by placing the barrel of the 105mm Field Howitzer M 18M
(Rheinmetall/Krupp) in the gun carriage of the 75mm Anti-Tank Gun M 40
(Rheinmetall). With this gun the Artillery of the German Infantry divisions
was to receive more effective protection from the tanks due to the much
lower silhouette.

Field Howitzer 105mm L

Country of origin: German Federal Republic
Manufacturer: Rheinmetall
Designation: 105mm L Light Field Howitzer
Calibre (mm): 105
Length of barrel (x calibre): 32
Depression/Elevation (°): -5 +65
Traverse (°): 45.5
Weight in firing position (kg): 2,500
Weight of projectile (kg): 15.0
Muzzle velocity (m/sec): 640
Maximum range (km): 14.17
Gun carriage: Split trail
Gun shield: Yes
Mode of traction: Motor drawn
Introduced into: GFR

Remarks:
Modification of the American 105mm Field Howitzer M-2A1 (Ordnance Department), with a longer barrel and other improvements.

140

Field Howitzer 120mm M 05   **Germany**

Country of origin: Germany
Manufacturer: Krupp
Designation: 12cm Field Howitzer Meiji 38
Calibre (mm): 120
Length of barrel (x calibre): 12
Depression/Elevation (°): -0 +42
Traverse (°): 5
Weight in firing position (kg): 1,125
Weight of projectile (kg): 20.0
Muzzle velocity (m/sec): 275
Maximum range (km): 5.8
Gun carriage: Box trail
Gun shield: No
Mode of traction: Horse drawn
Introduced into: Japan

Remarks:

Partly built in Japan. France partly supplied the material for the howitzers
to be built in Japan.

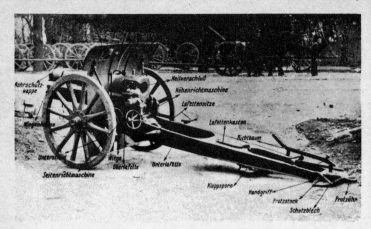

Country of origin: Germany
Manufacturer: Krupp
Designation: 12.2cm Howitzer M 09 (48 line Howitzer)
Calibre (mm): 121.9
Length of barrel (x calibre): 14
Depression/Elevation (°): -0 +43
Traverse (°): 4
Weight in firing position (kg): 1,350
Weight of projectile (kg): 23.0
Muzzle velocity (m/sec): 335
Maximum range (km): 7.5
Gun carriage: Box trail
Gun shield: Yes
Mode of traction: Horse drawn
Introduced into: Russia

Remarks:
Together with the 122mm Field Howitzer M 10 (Schneider), this was the
most important field howitzer at the beginning of World War I. Rebuilt in
the Soviet Union as the 122mm Field Howitzer M 09/37 (state).

Country of origin: Germany
Manufacturer: Krupp
Designation: 12cm Field Howitzer M 12
Calibre (mm): 120
Length of barrel (x calibre): 14
Depression/Elevation (^{O}): -11 +43
Traverse (^{O}): 3
Weight in firing position (kg): 1,420
Weight of projectile (kg): 21.0
Muzzle velocity (m/sec): 300
Maximum range (km): 6.0 *
Gun carriage: Box trail
Gun shield: Yes
Mode of traction: Horse or motor drawn
Introduced into: Switzerland

Remarks:
In use in the Field Howitzer Regiments of the army until the end of World
War I.

*Enlarged in 1938 by 1 km.

Country of origin: Germany
Manufacturer: Krupp
Designation: 12cm Light Howitzer L/14
Calibre (mm): 120
Length of barrel (x calibre): 14
Depression/Elevation (^{O}): -4.8 +43
Traverse (^{O}): 6
Weight in firing position (kg): 1,610
Weight of projectile (kg): 16.5
Muzzle velocity (m/sec): 317
Maximum range (km): 6.05
Gun carriage: Box trail
Gun shield: Yes
Mode of traction: Horse drawn
Introduced into: Netherlands

Remarks:

In use in the Heavy Artillery Regiments until the beginning of World War II.

Country of origin: Germany
Manufacturer: Rheinmetall
Designation: 12cm Field Howitzer M 09
Calibre (mm): 120
Length of barrel (x calibre): 13.5
Depression/Elevation (^{o}): -5 +43
Traverse (^{o}): 5
Weight in firing position (kg): 1,280
Weight of projectile (kg): 21.0
Muzzle velocity (m/sec): 300
Maximum range (km): 7.0
Gun carriage: Box trail
Gun shield: Yes
Mode of traction: Horse drawn
Introduced into: Norway

Remarks:

Country of origin: Germany/France
Manufacturer: Krupp/Hotchkiss
Designation: 3.7cm Mountain Gun L/23
Calibre (mm): 37
Length of barrel (x calibre): 23
Depression/Elevation (^{o}): -12 +28
Traverse (^{o}): 110/360 *
Weight in firing position (kg): 209
Weight of projectile (kg): 0.455
Muzzle velocity (m/sec): 400
Maximum range (km): 4.3
Gun carriage: Tripod/pedestal
Gun shield: Yes
Mode of traction: Dismantled
Introduced into: Germany

Remarks:

A modification of the captured French 37mm Infantry Gun M 16 (Hotchkiss). Used either on a pedestal platform or with a tripod.

* Tripod/pedestal

Country of origin: Germany
Manufacturer: Ehrhardt
Designation: 7.5cm Mountain Howitzer M 11
Calibre (mm): 75
Length of barrel (x calibre): 17
Depression/Elevation (°): -7 +39
Traverse (°): 6
Weight in firing position (kg): 530
Weight of projectile (kg): 5.3/6.5*
Muzzle velocity (m/sec): 350/315 *
Maximum range (km): 6.3/6.9*
Gun carriage: Box trail
Gun shield: Yes
Mode of traction: Horse drawn, pack animals
Introduced into: Norway

Remarks:
Dismantles into six parts. Nine batteries were supplied to Norway.

*With old/new shells.

Mountain Gun 75mm M 05 (M 04)　　　　　　　Germany

Country of origin: Germany
Manufacturer: Krupp
Designation: 7.5cm Mountain Gun M 05 (04)
Calibre (mm): 75
Length of barrel (x calibre): 14
Depression/Elevation (^{O}): -10 +25
Traverse (^{O}): 4
Weight in firing position (kg): 410/421*
Weight of projectile (kg): 5.1/5.3*
Muzzle velocity (m/sec): 330/300*
Maximum range (km): 5.0/4.8*
Gun carriage: Box trail
Gun shield: Yes
Mode of traction: Horse drawn, pack animals
Introduced into: Bulgaria (05), Turkey (04)

Remarks:

The picture shows M 05. Dismantles into four parts. 54 of these guns were supplied to Bulgaria from 1905 to 1907. Turkey had already ordered 23 batteries of 6 guns each in 1903.

*For M 05/M 04.

Country of origin: Germany
Manufacturer: Krupp
Designation: 7.5cm Mountain Gun M 1906
Calibre (mm): 75
Length of barrel (x calibre): 14
Depression/Elevation (°): -10 +25
Traverse (°): 4
Weight in firing position (kg): 409
Weight of projectile (kg): 5.3
Muzzle velocity (m/sec): 300
Maximum range (km): 3.6
Gun carriage: Box trail
Gun shield: Yes *
Mode of traction: Pack animals
Introduced into: Switzerland

Remarks:

Dismantles into four parts. Introduced after trials between 1902 and 1904.
Replaced by the 75mm Mountain Gun M 33 (Bofors).

* Added later.

Mountain Gun 75mm M 08 Germany

Country of origin: Germany
Manufacturer: Krupp
Designation: 7.5cm Mountain Gun Meiji 41
Calibre (mm): 75
Length of barrel (x calibre): 19.2
Depression/Elevation (°): -8 +25
Traverse (°): 7
Weight in firing position (kg): 544
Weight of projectile (kg): 6.8
Muzzle velocity (m/sec): 345
Maximum range (km): 5.1
Gun carriage: Pole trail
Gun shield: No
Mode of traction: Horse drawn, by pack animals or man power
Introduced into: Japan

Remarks:

Dismantles into six parts. Built mainly in Japan. Replaced the 75mm Mountain Gun M 98 with recoil carriage. Later rebuilt in Japan. Also used as an Infantry Gun after World War I.

Country of origin: Germany
Manufacturer: Krupp
Designation: 7.5cm Mountain Gun L/14
Calibre (mm): 75
Length of barrel (x calibre): 14
Depression/Elevation (°): -10 +30
Traverse (°): 5
Weight in firing position (kg): 550
Weight of projectile (kg): 5.3
Muzzle velocity (m/sec): 300
Maximum range (km): 5.4
Gun carriage: Box trail
Gun shield: Yes
Mode of traction: Horse drawn, by pack animals
Introduced into: Chile, Germany

Remarks:

Dismantles into six parts. Guns built for Chile. At the outbreak of World
War I a number of the existing guns were bought for Germany and they
were also built later.

Country of origin: Germany
Manufacturer: Rheinmetall
Designation: 7.5cm Mountain Gun L/17 M 08
Calibre (mm): 75
Length of barrel (x calibre): 17
Depression/Elevation (°): -7 +38.5
Traverse (°): 5
Weight in firing position (kg): 529
Weight of projectile (kg): 5.3
Muzzle velocity (m/sec): 300
Maximum range (km): 5.75
Gun carriage: Pole trail
Gun shield: Yes
Mode of traction: Horse drawn, by pack animals
Introduced into: Germany

Remarks:

Dismantles into five parts. The three batteries purchased were used only in the African Colonies.

Mountain Gun 75mm M 14

Country of origin: Germany
Manufacturer: Rheinmetall
Designation: 7.5cm Mountain Gun L/16 M 1914
Calibre (mm): 75
Length of barrel (x calibre): 16
Depression/Elevation (°): -7 +36
Traverse (°): 5
Weight in firing position (kg): 491
Weight of projectile (kg): 5.3
Muzzle velocity (m/sec): 280
Maximum range (km): 4.7
Gun carriage: Pole trail
Gun shield: Yes
Mode of traction: Horse drawn, by pack animals
Introduced into: China, Germany, Austria-Hungary, Turkey

Remarks:

Dismantles into five parts. Built for China. The 18 guns built by the outbreak of war were used by Germany. They were later given, with other newly built guns to Austria/Hungary and Turkey.

Country of origin: Germany
Manufacturer: Rheinmetall
Designation: 7.7cm Mountain Gun M 1915
Calibre (mm): 77
Length of barrel (x calibre): 17
Depression/Elevation (°): -7 +35
Traverse (°): 5
Weight in firing position (kg): 555
Weight of projectile (kg): 6.85
Muzzle velocity (m/sec): 310
Maximum range (km): 5.9
Gun carriage: Pole trail
Gun shield: Yes
Mode of traction: Horse drawn, by pack animals
Introduced into: Germany, Turkey

Remarks:

Dismantles into six parts. Replaced in Germany towards the end of World
War I by the 75mm Mountain Gun M 15 (Škoda). Seven batteries supplied
to Turkey. Also known as the 77mm Mountain Gun M 17.

Country of origin: Germany
Manufacturer: Rheinmetall
Designation: 7.5cm Mountain Gun 36
Calibre (mm): 75
Length of barrel (x calibre): 19.3
Depression/Elevation (°): -2 +70
Traverse (°): 40
Weight in firing position (kg): 750
Weight of projectile (kg): 5.8
Muzzle velocity (m/sec): 475
Maximum range (km): 9.25
Gun carriage: Split trail
Gun shield: No
Mode of traction: Horse and motor drawn, and by pack animals
Introduced into: Germany

Remarks:

Dismantles into eight parts. Replaced the old 75mm Mountain Gun M 15 (Škoda).

Country of origin: Germany
Manufacturer: Böhler
Designation: 10.5cm Mountain Howitzer 40
Calibre (mm): 105
Length of barrel (x calibre): 30
Depression/Elevation (°): -5.5 +71
Traverse (°): 51
Weight in firing position (kg): 1,660
Weight of projectile (kg): 14.5
Muzzle velocity (m/sec): 570
Maximum range (km): 12.625
Gun carriage: Split trail
Gun shield: No
Mode of traction: Horse and motor drawn
Introduced into: Germany

Remarks:

Dismantles into five parts.

Country of origin: Germany
Manufacturer: Krupp
Designation: 10.5cm Mountain Howitzer L/12
Calibre (mm): 105
Length of barrel (x calibre): 12
Depression/Elevation (°): -7 +40
Traverse (°): 5.5
Weight in firing position (kg): 845
Weight of projectile (kg): 14.4—15 8
Muzzle velocity (m/sec): 253
Maximum range (km): 4.9
Gun carriage: Box trail
Gun shield: Yes
Mode of traction: Horse drawn, by pack animals
Introduced into: Germany, Bulgaria, Turkey

Remarks:

Dismantles into eight parts. Designed to use the ammunition of the 105mm Field Howitzer M 98/09 (Krupp), except the 7 loading charge.

Country of origin: Great Britain
Manufacturer: Vickers
Designation: 7.62cm (13 Pdr) Fieldgun M 04
Calibre (mm): 76.2
Lenght of barrel (x calibre): 24.4
Depression/Elevation (°): -5 +16
Traverse (°): 8
Weight in firing position (kg): 998
Weight of projectile (kg): 5.9
Muzzle velocity (m/sec): 518
Maximum range (km): 7.6
Gun carriage: Pole trail
Gun shield: Yes
Mode of traction: Horse drawn
Introduced into: Great Britain

Remarks:
Used in the mounted batteries up until the beginning of World War I.
Developed in collaboration between Vickers, Sons & Maxim and
Armstrong, Whitworth and Co with the Woolwich Arsenal.

Country of origin: Great Britain
Manufacturer: Vickers
Designation: 8.38cm (18 Pdr) Fieldgun Mk I in Gun Carriage Mk I
Calibre (mm): 83.8
Length of barrel (x calibre): 29.4
Depression/Elevation (⁰): -5 +16
Traverse (⁰): 8
Weight in firing position (kg): 1,285
Weight of projectile (kg): 8.4
Muzzle velocity (m/sec): 492
Maximum range (km): 8.7/10.15
Gun carriage: Pole trail
Gun shield: Yes
Mode of traction: Horse drawn
Introduced into: Great Britain

Remarks:

Similar to Mk I, II and III in various gun carriages. Introduced into the army in 1904. The largest calibre fieldgun at the beginning of World War I. The range surpassed all other fieldguns. Developed in co-operation with the firms Vickers Sons & Maxim and Armstrong, Whitworth and Co with the Woolwich Arsenal.

Country of origin: Great Britain
Manufacturer: Vickers
Designation: 18 Pdr Fieldgun Mk IV in Gun Carriage Mk III
Calibre (mm): 83.8
Length of barrel (x calibre): 29.4
Depression/Elevation (°): -5 +30
Traverse (°): 9
Weight in firing position (kg): 1,514
Weight of projectile (kg): 8.4
Muzzle velocity (m/sec): 492
Maximum range (km): 10.15
Gun carriage: Box trail
Gun shield: Yes
Mode of traction: Horse or motor drawn
Introduced into: Great Britain, Estonia, Ireland, Latvia, China

Remarks:

Some had rubber tyres for motor traction. Some had a circular base plate.
Introduced after World War I. Still in use by the troops at the beginning of
World War II. Similar to the gun carriage of Mk IV.

Fieldgun 83.8mm Mk V   **Great Britain**

Country of origin: Great Britain
Manufacturer: Vickers
Designation: 18 Pdr Fieldgun Mk IV—in Gun Carriage Mk V
Calibre (mm): 83.8
Length of barrel (x calibre): 29.4
Depression/Elevation (°): -4.7 +37.8
Traverse (°): 50
Weight in firing position (kg): 1,593
Weight of projectile (kg): 8.4
Muzzle velocity (m/sec): 505
Maximum range (km): 10.15
Gun carriage: Split trail
Gun shield: Yes
Mode of traction: Horse or motor drawn
Introduced into: Great Britain

Remarks:

With pneumatic tyres for motor traction. Introduced after World War I to replace the old model Mk I-IV.

Country of origin: Great Britain
Manufacturer: state/Vickers
Designation: 25 Pdr Gun Howitzer Mk I in Gun Carriage Mk IV P
Calibre (mm): 87.6
Length of barrel (x calibre): 28.1
Depression/Elevation (^{o}): -4 +30
Traverse (^{o}): 9
Weight in firing position (kg): 1,490
Weight of projectile (kg): 11.34
Muzzle velocity (m/sec): 520
Maximum range (km): 10.8
Gun carriage: Box trail
Gun shield: Yes
Mode of traction: Motor drawn
Introduced into: Great Britain, Ireland

Remarks:

Bored out version of the 83.8mm Fieldgun Mk II-IV (Vickers) and smaller
numbers also of the 83.8mm Fieldgun Mk V (Vickers) with split trail gun
carriage. Replaced in Britain by the 87.6mm Gun Howitzer Mk II (state).

Gun Howitzer 87.6mm Mk II — Great Britain

Country of origin: Great Britain
Manufacturer: state
Designation: 25 Pdr Gun Howitzer Mk II in Gun Carriage Mk I
Calibre (mm): 87.6
Length of barrel (x calibre): 28.3
Depression/Elevation (^{o}): -5 +42
Traverse (^{o}): 8/360 *
Weight in firing position (kg): 1,742
Weight of projectile (kg): 11.34
Muzzle velocity (m/sec): 532
Maximum range (km): 12.25
Gun carriage: Box trail
Gun shield: Yes
Mode of traction: Motor drawn
Introduced into: Great Britain, Denmark, Italy, Netherlands

Remarks:

Introduced in 1935/36 after long trials, and remained in use after 1940. Fitted with muzzle brake in 1943 and used as an anti-tank weapon. Later replaced by self-propelled guns and the Italian 105mm Mountain Howitzer M 56 (OTO Melara).

* Without/with circular baseplate.

Country of origin: Great Britain
Manufacturer: Vickers
Designation: 10.5cm Field Howitzer M 22
Calibre (mm): 105
Length of barrel (x calibre): 23.6
Depression/Elevation (°): -5 +37.5
Traverse (°): 9
Weight in firing position (kg): 1,577
Weight of projectile (kg): 12.00
Muzzle velocity (m/sec): 457
Maximum range (km): 12.0
Gun carriage: Box trail
Gun shield: Yes
Mode of traction: Horse drawn
Introduced into: Spain

Remarks:

Country of origin: Great Britain
Manufacturer: Vickers
Designation: 4.5'' Howitzer Mk I
Calibre (mm): 114.3
Length of barrel (x calibre): 15.6
Depression/Elevation (°): -5 +45
Traverse (°): 6
Weight in firing position (kg): 1,370
Weight of projectile (kg): 15.9
Muzzle velocity (m/sec): 313
Maximum range (km): 6.4/7.5 *
Gun carriage: Box trail
Gun shield: Yes
Mode of traction: Horse drawn
Introduced into: Great Britain, Ireland, Poland, Portugal, Rumania, Russia

Remarks:

Also made in Coventry. Introduced 1910. At the beginning of World War I
this gun represented one of the most efficient field howitzers. In the
autumn of 1918 there were 1,225 in service.

*Original/later increase.

Country of origin: Great Britain
Manufacturer: Vickers
Designation: 4.5″ Howitzer Mk II
Calibre (mm): 114.3
Length of barrel (x calibre): 15.6
Depression/Elevation (^{O}): -5 +45
Traverse (^{O}): 6
Weight in firing position (kg): 1,475
Weight of projectile (kg): 15.9
Muzzle velocity (m/sec): 313
Maximum range (km): 7.5
Gun carriage: Box trail
Gun shield: Yes
Mode of traction: Horse or motor drawn
Introduced into: Great Britain

Remarks:

Use by the Brigades of the Field Artillery until 1944. Replaced by the 87.6mm Gun Howitzer Mk II (state).

Country of origin: Great Britain
Manufacturer:
Designation: 5'' Howitzer Mk II
Calibre (mm): 127
Length of barrel (x calibre): 9.8
Depression/Elevation (^{O}): -5 +45
Traverse (^{O}): 0
Weight in firing position (kg): 1,169
Weight of projectile (kg): 18.2/22.7
Muzzle velocity (m/sec): -/238
Maximum range (km): 6.0/4.5
Gun carriage: Box trail
Gun shield: No
Mode of traction: Horse drawn
Introduced into: Great Britain, Russia

Remarks:
Replaced by the 114mm Field Howitzer M 10 (Vickers)

Country of origin: Great Britain
Manufacturer: Vickers
Designation: 2.75″ Mountain Gun Mk I
Calibre (mm): 69.8
Length of barrel (x calibre): 27.8
Depression/Elevation (°): -15 +22
Traverse (°): 8
Weight in firing position (kg): 586
Weight of projectile (kg): 5.67
Muzzle velocity (m/sec): 393
Maximum range (km): 5.4
Gun carriage: Box trail
Gun shield: No
Mode of traction: Horse drawn, by pack animals
Introduced into: Great Britain

Remarks:
Dismantles into six parts. Primarily intended for use in the Colonies, (Tibet, Aden).

Country of origin: Great Britain
Manufacturer: Vickers
Designation: 7.5cm (2.95'') Mountain Gun MI
Calibre (mm): 75
Length of barrel (x calibre): 12.1
Depression/Elevation (°): -10 +27
Traverse (°): —
Weight in firing position (kg): 376
Weight of projectile (kg): 5.67—8.17
Muzzle velocity (m/sec): 280—230
Maximum range (km): 4.5
Gun carriage: Box trail
Gun shield: No
Mode of traction: By pack animals
Introduced into: Great Britain, USA

Remarks:

Dismantles into four parts. The Maxim-Nordenfelt System. Great Britain intended this mainly for use in the Colonies (Sudan campaign). Replaced in the USA by the 75mm Mountain Howitzer M 1 (Ordnance Department).

Country of origin: Great Britain
Manufacturer: Vickers
Designation: 9.4cm (3.7'') Mountain Howitzer Mk I
Calibre (mm): 93.9
Length of barrel (x calibre): 12.6
Depression/Elevation (^{O}): -5 +40
Traverse (^{O}): 40
Weight in firing position (kg): 779
Weight of projectile (kg): 9.07
Muzzle velocity (m/sec): 294
Maximum range (km): 5.5
Gun carriage: Split trail
Gun shield: Yes
Mode of traction: Horse and motor drawn, by pack animals
Introduced into: Great Britain

Remarks:

Dismantles into eight parts. Developed during World War I. Later used as an Infantry gun and some had rubber tyres for motor traction. Barrel of gun can be dismantled.

Fieldgun 105mm Great Britain

Country of origin: Great Britain
Manufacturer: Royal Armaments Research and Development Establishment
Designation: 105mm Light Fieldgun
Calibre (mm): 105
Length of barrel (x calibre): —
Depression/Elevation (°): -5.5 +70
Traverse (°): 11/360 *
Weight in firing position (kg): 1,768
Weight of projectile (kg): 15
Muzzle velocity (m/sec): —
Maximum range (km): 15.0/17.4 †
Gun carriage: Pole trail
Gun shield: No
Mode of traction: Motor drawn
Introduced into: Great Britain

Remarks:
Intended to replace the present 87.6mm Gun Howitzer Mk II (state) in 1974.

* On gun carriage/on baseplate.
† With super charge.

171

Country of origin: Hungary
Manufacturer: MÁVAG
Designation: 10.5cm M 40 Field Howitzer
Calibre (mm): 105
Length of barrel (x calibre): 20.5
Depression/Elevation (°): —
Traverse (°): —
Weight in firing position (kg): 1,600
Weight of projectile (kg): 15.0/17.0
Muzzle velocity (m/sec): 448/—
Maximum range (km): 10.4
Gun carriage: —
Gun shield: Yes
Mode of traction: Horse drawn
Introduced into: Hungary

Remarks:
Only built in small numbers. Used mainly as armament for the Assault Howitzer 'Zrinyi'.

Country of origin: Hungary/Austria-Hungary
Manufacturer: MÁVAG/Škoda
Designation: 7.5cm M 15/35 Mountain Gun
Calibre (mm): 75
Length of barrel (x calibre): 15
Depression/Elevation (^{0}): -9 +50
Traverse (^{0}): 7
Weight in firing position (kg): —
Weight of projectile (kg): 6.5
Muzzle velocity (m/sec): —
Maximum range (km): 7.6
Gun carriage: Box trail
Gun shield: Yes
Mode of traction: Horse drawn
Introduced into: Hungary

Remarks:

Similar to the 75mm Mountain Gun M 15/31. Hungarian modification of the 75mm Mountain Gun M 15 (Škoda) to be horse drawn.

Country of origin: Italy
Manufacturer: Ansaldo
Designation: Gun 75/32 M 37
Calibre (mm): 75
Length of barrel (x calibre): 32
Depression/Elevation (°): -10 + 19.5
Traverse (°): 50
Weight in firing position (kg): 1,200
Weight of projectile (kg): 6.35
Muzzle velocity (m/sec): 624
Maximum range (km): 12.58
Gun carriage: Split trail
Gun shield: Yes
Mode of traction: Motor drawn
Introduced into: Italy

Remarks:
A small number only were built. Intended for use by Artillery Regiments of
the 'Fast Divisions'.

Country of origin: Italy
Manufacturer: Ansaldo
Designation: Howitzer 75/18 M 35
Calibre (mm): 75
Length of barrel (x calibre): 18.3
Depression/Elevation (^{o}): -10 +45
Traverse (^{o}): 50
Weight in firing position (kg): 1,065
Weight of projectile (kg): 6.35
Muzzle velocity (m/sec): 435
Maximum range (km): 9.4
Gun carriage: Split trail
Gun shield: Yes
Mode of traction: Horse and motor drawn, in sections by the troops
Introduced into: Italy

Remarks:

In two sections tandem drawn for mountain use. Dismantles further for transportation by the troops. Ballistically the same as the 75mm Mountain Gun M 34. Introduced mainly for mounted and mobile (motorised) batteries.

Country of origin: Italy
Manufacturer: Arsenal Turin
Designation: Mountain Gun 65/17
Calibre (mm): 65
Length of barrel (x calibre): 17
Depression/Elevation (°): -10 +20
Traverse (°): 8
Weight in firing position (kg): 556
Weight of projectile (kg): 4.3—4.92
Muzzle velocity (m/sec): 345
Maximum range (km): 6.8
Gun carriage: Box trail
Gun shield: No
Mode of traction: Horse drawn, by pack animals
Introduced into: Italy

Remarks:

Dismantles into six parts. Introduced before World War I. After the war replaced by the captured Austro--Hungarian 75mm Mountain Gun M 15 (Škoda). Later used as an Infantry gun. At the beginning of World War II there were still 700 of these guns in existence.

Country of origin: Italy
Manufacturer: Arsenal Turin
Designation: Mountain Gun 70 A
Calibre (mm): 70
Length of barrel (x calibre): 16.4
Depression/Elevation (^{O}): -12 +21
Traverse (^{O}): —
Weight in firing position (kg): 387
Weight of projectile (kg): 4.84
Muzzle velocity (m/sec): 353
Maximum range (km): 6.63
Gun carriage: Box trail
Gun shield: No
Mode of traction: Horse drawn, by pack animals
Introduced into: Italy

Remarks:

Dismantles into four parts. With rope and trail spade. Introduced 1904.
Replaced by the 65mm Mountain Gun M 13 (Arsenal Turin).

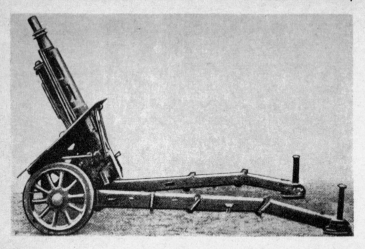

Country of origin: Italy
Manufacturer: Ansaldo
Designation: Mountain Gun 75/18 M 34
Calibre (mm): 75
Length of barrel (x calibre): 18.3
Depression/Elevation (^{o}): -10 +65
Traverse (^{o}): 50
Weight in firing position (kg): 800
Weight of projectile (kg): 6.35
Muzzle velocity (m/sec): 435
Maximum range (km): 9.4
Gun carriage: Split trail
Gun shield: Yes
Mode of traction: Horse and motor drawn, by pack animals
Introduced into: Italy

Remarks:

Dismantles into eight parts. Ballistically the same as the 75mm Field Howitzer M 35 (Ansaldo). Replaced the old 65mm Mountain Gun M 13 (Arsenal Turin) and the 75mm Mountain Gun M 15 (Škoda).

Country of origin: Italy
Manufacturer: OTO Melara
Designation: 105mm Light Mountain Howitzer 105/14
Calibre (mm): 105
Length of barrel (x calibre): 14
Depression/Elevation (°): -7 +65
Traverse (°): 36 or 56
Weight in firing position (kg): 1,273
Weight of projectile (kg): 14.97
Muzzle velocity (m/sec): 416
Maximum range (km): 11.1
Gun carriage: Split trail
Gun shield: Yes
Mode of traction: Motor drawn, pack animals, manpower
Introduced into: Italy, GFR, Great Britain, Spain, France, Belgium

Remarks:

Dismantles into twelve parts. Developed after World War II. Uses the
same ammunition as the American 105mm Field Howitzer M-2A1
(Ordnance Department). Gun carriage can be raised and lowered.

Country of origin: Japan
Manufacturer: Arisaka/Osaka Arsenal
Designation: 7.5cm Fieldgun Meiji 38
Calibre (mm): 75
Length of barrel (x calibre): 30
Depression/Elevation (°): -8 +20
Traverse (°): 7
Weight in firing position (kg): 980
Weight of projectile (kg): 6.5/5.7—7.0 *
Muzzle velocity (m/sec): 510/— *
maximum range (km): 8.3/10.0 *
Gun carriage: Box trail
Gun shield: Yes
Mode of traction: Horse drawn
Introduced into: Japan

Remarks:

Modified version of the 75mm Fieldgun M 05 (Krupp), with improvements by General Arisaka. Produced at the Osaka Arsenal, where some of these guns were later rebuilt as the 75mm Fieldgun M 05 (improved).

*With old/new projectile M 30.

Country of origin: Japan/Germany
Manufacturer: Osaka Arsenal/Krupp
Designation: 7.5cm Fieldgun Meiji 41
Calibre (mm): 75
Length of barrel (x calibre): 30
Depression/Elevation (°): -8 +30
Traverse (°): 7
Weight in firing position (kg): 980
Weight of projectile (kg): 6.5/5.7—7.0 *
Muzzle velocity (m/sec): 520/— *
Maximum range (km): 8.3/—*
Gun carriage: Box trail
Gun shield: Yes
Mode of traction: Horse drawn
Introduced into: Japan

Remarks:

Built in Japan under licence from Krupp. Introduced into the Artillery units
of the Cavalry divisions. Later partly rebuilt.

* With old/new projectile.

Country of origin: Japan
Manufacturer: Osaka Arsenal
Designation: 7.5cm Fieldgun Meiji 38 (improved)
Calibre (mm): 75
Length of barrel (x calibre): 31
Depression/Elevation (º): -8 +43
Traverse (º): 7
Weight in firing position (kg): 1,130
Weight of projectile (kg): 6.5/5.7 – 7.0 *
Muzzle velocity (m/sec): 520/ – *
Maximum range (km): 8.3/10.0/11.5 *
Gun carriage: Box trail
Gun shield: Yes
Mode of traction: Horse drawn
Introduced into: Japan

Remarks:

Improved version of the 75mm Fieldgun M 05 (Arisaka/Osaka Arsenal).
The design was based on that of the 75mm Fieldgun M 05 (Krupp). Partly
replaced the two older models. It was the most widely used gun in Japan
during World War II.

* With old/new projectiles M 05/M 30.

Country of origin: Japan/France (?)
Manufacturer: Osaka Arsenal/Schneider (?)
Designation: 7.5cm Fieldgun M 90
Calibre (mm): 75
Length of barrel (x calibre): 38
Depression/Elevation (°): +43
Traverse (°): 50
Weight in firing position (kg): 1,397
Weigh of projectile (kg): 5.7−7.0
Muzzle velocity (m/sec): 700
Maximum range (km): 14.0
Gun carriage: Split trail
Gun shield: Yes
Mode of traction: Horse and motor drawn
Introduced into: Japan

Remarks:

Built in Japan. Design by Schneider (?). From 1936 partly replaced the old
fieldguns in the Artillery regiments of the Divisions.

Country of origin: Japan
Manufacturer: Osaka Arsenal
Designation: 7.5cm Fieldgun M 95
Calibre (mm): 75
Length of barrel (x calibre): 31
Depression/Elevation (^{o}): -8 +43
Traverse (^{o}): 50
Weight in firing position (kg): 1,007
Weight of projectile (kg): 5.7 − 7.0
Muzzle velocity (m/sec): 520
Maximum range (km): 10.7
Gun carriage: Split trail
Gun shield: Yes
Mode of traction: Horse drawn
Introduced into: Japan

Remarks:
Introduced as a replacement for the obsolete 75mm Fieldgun M 08 (Osaka Arsenal/Krupp) of the mounted batteries.

Country of origin: Japan
Manufacturer: Osaka Arsenal
Designation: 10.5cm Field Howitzer M 91
Calibre (mm): 105
Length of barrel (x calibre): 24
Depression/Elevation (^{O}): -5 +45
Traverse (^{O}): 40
Weight in firing position (kg): 1,497
Weight of projectile (kg): 15.8
Muzzle velocity (m/sec): 545
Maximum range (km): 10.8
Gun carriage: Split trail
Gun shield: Yes
Mode of traction: Horse drawn
Introduced into: Japan

Remarks:

Design by Schneider (?) Replaced the obsolete 120mm Field Howitzer M 05 (Krupp).

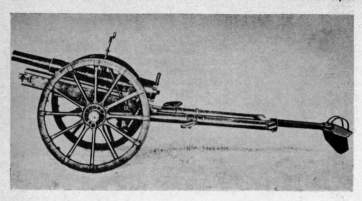

Country of origin: Japan
Manufacturer: Osaka Arsenal
Designation: 7.5cm Mountain Gun Meiji 41
Calibre (mm): 75
Length of barrel (x calibre): 19.2
Depression/Elevation (°): -18 +40
Traverse (°): 6
Weight in firing position (kg): 544
Weight of projectile (kg): 6.02—7.0
Muzzle velocity (m/sec): 435
Maximum range (km): 7.1
Gun carriage: Pole trail
Gun shield: Yes
Mode of traction: Horse drawn, by the troops and by pack animals
Introduced into: Japan

Remarks:

Dismantles into six parts. Modified version of the 75mm Mountain Gun M 08 (Krupp). Used in World War II as an Infantry gun. The first Japanese gun to fire hollow charge shells. Replaced by the 75mm Mountain Gun M 34 (Osaka Arsenal).

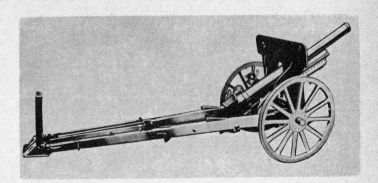

Country of origin: Japan
Manufacturer: Osaka Arsenal
Designation: 7.5cm Mountain Gun M 94
Calibre (mm): 75
Length of barrel (x calibre): 21
Depression/Elevation (°): -9 +45
Traverse (°): 40
Weight in firing position (kg): 544
Weight of projectile (kg): 5.7/6.02—7.0
Muzzle velocity (m/sec): —/355
Maximum range (km): 8.0/7.0
Gun carriage: Split trail
Gun shield: Yes
Mode of traction: Horse drawn, by pack animals
Introduced into: Japan

Remarks:

Dismantles into eleven parts for six pack animals. Replaced the 75mm Mountain Guns M 08 (Krupp) and M 08 mod (Osaka Arsenal) in the units of the Mountain Artillery.

Fieldgun 75mm M 02/04 vd OM　　　　　　　The Netherlands

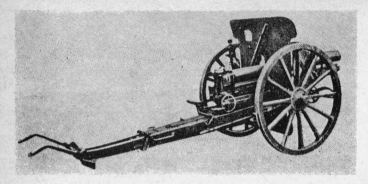

Country of origin: The Netherlands/Germany
Manufacturer: Siderius/Krupp
Designation: 7.5cm Fieldgun (modernised) OM
Calibre (mm): 75
Length of barrel (x calibre): 30
Depression/Elevation (^{O}): -10 +16
Traverse (^{O}): 7
Weight in firing position (kg): 980
Weight of projectile (kg): 6.5
Muzzle velocity (m/sec): 500
Maximum range (km): 6.9
Gun carriage: Box trail
Gun shield: Yes
Mode of traction: Horse drawn
Introduced into: The Netherlands

Remarks:

A Krupp gun modernised in The Netherlands. The 75mm Fieldgun M 03 (Krupp) rebuilt after World War I.

Country of origin: The Netherlands/Germany
Manufacturer: Siderius/Krupp
Designation: 7.5cm Fieldgun (modernised)
Calibre (mm): 75
Length of barrel (x calibre): 30
Depression/Elevation (^{0}): —
Traverse (^{0}): —
Weight in firing position (kg): —
Weight of projectile (kg): 6.5
Muzzle velocity (m/sec): 500
Maximum range (km): 8.7
Gun carriage: Box trail
Gun shield: Yes
Mode of traction: Horse drawn
Introduced into: The Netherlands

Remarks:

A Krupp gun modernised in The Netherlands. The 75mm Fieldgun M 03
(Krupp) rebuilt after World War I.

Country of origin: The Netherlands/Germany
Manufacturer: Siderius/Krupp
Designation: 7.5cm Fieldgun (modernised) NM
Calibre (mm): 75
Length of barrel (x calibre): 30
Depression/Elevation (^{O}): -8 +40
Traverse (^{O}): 9
Weight in firing position (kg): 1,299
Weight of projectile (kg): 6.5
Muzzle velocity (m/sec): 544
Maximum range (km): 10.6
Gun carriage: Box trail
Gun shield: Yes
Mode of traction: Horse drawn
Introduced into: The Netherlands

Remarks:

A Krupp gun modernised in The Netherlands. The 75mm Fieldgun M 03 (Krupp) rebuilt after World War I.

Country of origin: Norway
Manufacturer: Kongsberg
Designation: 12cm Field Howitzer
Calibre (mm): 120
Length of barrel (x calibre): 20
Depression/Elevation (^{o}): -5 +45
Traverse (^{o}): 54
Weight in firing position (kg): 1,970
Weight of projectile (kg): 20.4
Muzzle velocity (m/sec): 450
Maximum range (km): 10.3
Gun carriage: Split trail
Gun shield: Yes
Mode of traction: Horse or motor drawn
Introduced into: Norway

Remarks:

Country of origin: Norway
Manufacturer: Kongsberg
Designation: 7.5cm Mountain Gun M 27
Calibre (mm): 75
Length of barrel (x calibre): 20.5
Depression/Elevation (^{o}): -5 +47
Traverse (^{o}): 5
Weight in firing position (kg): 600
Weight of projectile (kg): 6.5
Muzzle velocity (m/sec): 395
Maximum range (km): 8.8
Gun carriage: Box trail
Gun shield: —
Mode of traction: Horse drawn, by pack animals
Introduced into: Norway

Remarks:

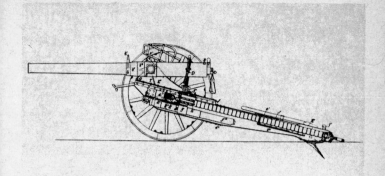

Country of origin: Russia
Manufacturer: Putilow
Designation: 7.62cm Fieldgun M 00
Calibre (mm): 76.2
Length of barrel (x calibre): 30
Depression/Elevation (^{O}): -6.5 +16.75
Traverse (^{O}): 4
Weight in firing position (kg): 1,020*
Weight of projectile (kg): 6.50
Muzzle velocity (m/sec): 588
Maximum range (km): 6.4
Gun carriage: Box trail
Gun shield: No †
Mode of traction: Horse drawn
Introduced into: Russia

Remarks:

About 1,200 guns were built. The main weapon of the Russian artillery in the Manchurian War. Replaced by the 76.2mm Fieldgun M 02 (Putilow).

* 970kg for mounted batteries.
† Shield added about 1903.

Fieldgun 76.2mm M 02*

Country of origin: Russia
Manufacturer: Putilow
Designation: 7.62cm Fieldgun M 02
Calibre (mm): 76.2
Length of barrel (x calibre): 30
Depression/Elevation (^{O}): -6 +17
Traverse (^{O}): 5.5
Weight in firing position (kg): 1,040 †
Weight of projectile (kg): 6.58/7.45 ‡
Muzzle velocity)m/sec): 593/588 ‡
Maximum range (km): 6.6/8.75 ‡
Gun carriage: Box trail
Gun shield: Yes
Mode of traction: Horse drawn
Introduced into: Russia, Finland, Lithuania, Rumania

Remarks:

Introduced after the Manchurian War as a replacement for the 76.2mm
Fieldgun M 00 (Putilow).

* Sometimes known as the M 00/02
† 975kg for mounted batteries.
‡ With old/new shells.

Country of origin: Soviet Union/Russia
Manufacturer: state/Putilow
Designation: 7.5cm Fieldgun M 02/26
Calibre (mm): 75
Length of barrel (x calibre): 30
Depression/Elevation (^{O}): -6 +16
Traverse (^{O}): 5
Weight in firing position (kg): 1,040
Weight of projectile (kg): 6.5
Muzzle velocity (m/sec): 593
Maximum range (km): 8.5
Gun carriage: Box trail
Gun shield: Yes
Mode of traction: Horse drawn
Introduced into: Poland

Remarks:
Rebuild of the old 76.2mm Fieldgun M 02 (Putilow). Used by the mounted batteries.

Country of origin: Soviet Union/Russia
Manufacturer: state/Putilow
Designation: 76mm Division Gun 02/30
Calibre (mm): 76.2
Length of barrel (x calibre): 40
Depression/Elevation (^{o}): -5 +37
Traverse (^{o}): 5
Weight in firing position (kg): 1,350
Weight of projectile (kg): 6.4
Muzzle velocity (m/sec): 680
Maximum range (km): 13.0
Gun carriage: Box trail
Gun shield: Yes
Mode of traction: Horse drawn
Introduced into: Soviet Union

Remarks:

Improvement of the old 76.2mm Fieldgun M 02 (Putilow), by increasing
the vertical range and also partly by developing a longer barrel. Also
known under the same designation was a gun L/30, muzzle velocity 635
m/sec, range 12.4 km, weight 1320 kg.

Fieldgun 76.2mm M 33

Country of origin: Soviet Union
Manufacturer: state
Designation: 76mm Division Gun 33
Calibre (mm): 76.2
Length of barrel (x calibre): 50
Depression/Elevation (°): -3 +43
Traverse (°): 4.7
Weight in firing position (kg): 1,600
Weight of projectile (kg): 6.4
Muzzle velocity (m/sec): 706
Maximum range (km): 13.6
Gun carriage: Box trail
Gun shield: Yes
Mode of traction: Horse drawn
Introduced into: Soviet Union

Remarks:

Used in the gun carriage of the 122mm Field Howitzer M 10/30 (state/Schneider).

Country of origin: Soviet Union
Manufacturer: state
Designation: 76mm Division Gun 36
Calibre (mm): 76.2
Length of barrel (x calibre): 51
Depression/Elevation (^{o}): -5 +75
Traverse (^{o}): 60
Weight in firing position (kg): 1,620
Weight of projectile (kg): 6.4
Muzzle velocity (m/sec): 706
Maximum range (km): 13.6
Gun carriage: Split trail
Gun shield: Yes
Mode of traction: Motor drawn
Introduced into: Soviet Union, Finland

Remarks:
Guns captured in Germany and modified for use as anti-tank guns. The
gun was intended for use as a fieldgun and as an anti-aircraft gun.

Country of origin: Soviet Union
Manufacturer: state
Designation: 76mm Division Gun 39
Calibre (mm): 76
Length of barrel (x calibre): 42
Depression/Elevation (°): -6 +45
Traverse (°): 57
Weight in firing position (kg): 1,484
Weight of projectile (kg): 6.4
Muzzle velocity (m/sec): 680
Maximum range: 13.3
Gun carriage: Split trail
Gun shield: Yes
Mode of traction: Motor drawn
Introduced into: Soviet Union

Remarks:

Guns captured and modified in Germany for use as anti-tank guns.

Country of origin: Soviet Union
Manufacturer: state
Designation: 76mm Gun 41
Calibre (mm): 76.2
Length of barrel (x calibre): 42.6
Depression/Elevation (^{o}): -8 +18
Traverse (^{o}): 56
Weight in firing position (kg): 1,110
Weight of projectile (kg): 6.2
Muzzle velocity (m/sec): 680
Maximum range (km): 13.0
Gun carriage: Split trail
Gun carriage: Yes
Mode of traction: Motor drawn
Introduced into: Soviet Union

Remarks:
Standard gun carriage with Anti-Tank Gun 57mm M 41. Only produced in
small numbers. With the barrel in a new gun carriage it became known as
the 76.2mm Fieldgun M 42 (state).

Fieldgun 76.2mm M 39/42

Soviet Union

Country of origin: Soviet Union
Manufacturer: state
Designation: 76mm Division Gun 39/42
Calibre (mm): 76.2
Length of barrel (x calibre): 42
Depression/Elevation (^{o}): -6 +45
Traverse (^{o}): 57
Weight in firing position (kg): 1,596
Weight of projectile (kg): 6.4
Muzzle velocity (m/sec): 680
Maximum range (km): 13.3
Gun carriage: Split trail
Gun shield: Yes
Mode of traction: Motor drawn
Introduced into: Soviet Union

Remarks:

Barrel of the 76.2mm Fieldgun M 39 (state) in a new gun carriage. Replaced by the 76.2mm Fieldgun M 42 (state/SIS).

Country of origin: Soviet Union
Manufacturer: state/SIS
Designation: 76mm Gun 42
Calibre (mm): 76.2
Length of barrel (x calibre): 42.6
Depression/Elevation (°): -5 +37
Traverse (°): 54
Weight in firing position (kg): 1,116
Weight of projectile (kg): 6.2
Muzzle velocity (m/sec): 680
Maximum range (km): 13.3
Gun carriage: Split trail (tubular)
Gun shield: Yes
Mode of traction: Motor drawn
Introduced into: Soviet Union, Eastern Bloc, Austria

Remarks:
A combination of fieldgun and anti-tank gun. Used by all states in the
Eastern Bloc after World War II. Barrel of the 76.2mm Fieldgun M 41
(state) in a new gun carriage.

Country of origin: Soviet Union
Manufacturer: state
Designation: 85mm Division Gun 43
Calibre (mm): 85
Length of barrel (x calibre): 55
Depression/Elevation (^{o}): +40
Traverse (^{o}): 30
Weight in firing position (kg): 1,704
Weight of projectile (kg): 9.5
Muzzle velocity (m/sec): 793
Maximum range (km): 16.6
Gun carriage: Split trail (tubular)
Gun shield: Yes
Mode of traction: Motor drawn
Introduced into: Soviet Union

Remarks:
Used mainly as an anti-tank gun.

Country of origin: Soviet Union
Manufacturer: state
Designation: 85mm Division Gun 45 *
Calibre (mm): 85
Length of barrel (x calibre): 55
Depression/Elevation (°): -5 +35
Traverse (°): 54
Weight in firing position (kg): 1,725
Weight of projectile (kg): 9.5
Muzzle velocity (m/sec): 793
Maximum range (km): 15.5
Gun carriage: Split trail (tubular)
Gun shield: Yes
Mode of traction: Motor drawn
Introduced into: Soviet Union, Eastern Bloc

Remarks:
Used mainly as an anti-tank gun.

* Also known as the M 44.

Country of origin: Soviet Union
Manufacturer: state
Designation: 85mm Gun 45 (HA)
Calibre (mm): 85
Length of barrel (x calibre): 55
Depression/Elevation (°): -5 +40
Traverse (°): 54
Weight in firing position (kg): 2,100
Weight of projectile (kg): 9.5
Muzzle velocity (m/sec): 793
Maximum range (km): 16.0
Gun carriage: Split trail (tubular)
Gun shield: Yes
Mode of traction: Motor drawn, self propelled
Introduced into: Soviet Union, Eastern Bloc

Remarks:

Barrel and carriage of the 85mm Fieldgun M 45 (state), with auxiliary drive.
A 750ccm motor cycle engine is built on a transom, and this engine drives
the wheels of the main axle. It has a steerable spur wheel and a driver's
seat. Used mainly as an anti-tank weapon. Designated Self Propelled Gun
SFK 85 by the East German Army.

Country of origin: Russia
Manufacturer: Putilow/Obuchow
Designation: 48 Line Howitzer M 04
Calibre (mm): 121.9
Length of barrel (x calibre): 12
Depression/Elevation (^{o}): -0 +42
Traverse (^{o}): 4.33
Weight in firing position (kg): 1,225
Weight of projectile (kg): 21.0
Muzzle velocity (m/sec): 292
Maximum range (km): 6.7
Gun carriage: Box trail
Gun shield: No
Mode of traction: Horse drawn
Introduced into: Russia

Remarks:

Country of origin: Soviet Union/France
Manufacturer: state/Schneider
Designation: 122mm Howitzer M 10/30
Calibre (mm): 121.9
Length of barrel (x calibre): 12.8
Depression/Elevation (°): -3 +43
Traverse (°): 4.7
Weight in firing position (kg): 1,465
Weight of projectile (kg): 21.76
Muzzle velocity (m/sec): 364
Maximum range (km): 8.9
Gun carriage: Box trail
Gun shield: Yes
Mode of traction: Horse drawn
Introduced into: Soviet Union

Remarks:

Improvement of the 122mm Field Howitzer M 10 (Schneider). The strengthened gun carriage made it possible to increase the muzzle velocity using the old barrel.

Country of origin: Soviet Union/Germany
Manufacturer: state/Krupp
Designation: 122mm Howitzer M 09/37
Calibre (mm): 121.9
Length of barrel (x calibre): 14
Depression/Elevation (^{O}): -1 +43
Traverse (^{O}): 4
Weight in firing position (kg): 1,450
Weight of projectile (kg): 21.76
Muzzle velocity (m/sec): 364
Maximum range (km): 8.9
Gun carriage: Box trail
Gun shield: Yes
Mode of traction: Horse drawn
Introduced into: Soviet Union

Remarks:

Improved version of the 122mm Field Howitzer M 09 (Krupp). The strengthened gun carriage made it possible to increase the muzzle velocity using the old barrel.

Country of origin: Soviet Union
Manufacturer: state
Designation: 122mm Howitzer 38
Calibre (mm): 121.9
Length of barrel (x calibre): 22.7
Depression/Elevation (°): -3 +63.5
Traverse (°): 49
Weight in firing position (kg): 2,450
Weight of projectile (kg): 21.76
Muzzle velocity (m/sec): 515
Maximum range (km): 11.8
Gun carriage: Split trail
Gun shield: Yes
Mode of traction: Motor drawn
Introduced into: Soviet Union, Eastern Bloc, Finland

Remarks:

Standard gun carriage with the 152mm Howitzer M 43. Used in World War
II by the divisional artillery and the Army Group Artillery. After the war it
was the standard howitzer of the Soviet Artillery. In 1965 the gun was
replaced in the divisions of the First Strategic Echelon by the 122mm Gun
Howitzer M 63 (state).

Country of origin: Soviet Union
Manufacturer: state
Designation: 122mm Gun Howitzer 63
Calibre (mm): 121.9
Length of barrel (x calibre): —
Depression/Elevation (^{o}): —
Traverse (^{o}): 360
Weight in firing position (kg): 5,000 (3,200?)
Weight of projectile (kg): —
Muzzle velocity (m/sec): 690
Maximum range (km): 17.0
Gun carriage: Triple spar
Gun shield: Yes
Mode of traction: Motor drawn
Introduced into: Soviet Union, partly by the Warsaw Pact

Remarks:

Replaced the 122mm Field Howitzer M 38 (state). Was to be the standard
weapon of the divisional artillery. Also used as an anti-tank weapon.

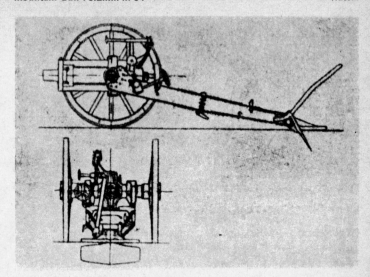

Country of origin: Russia
Manufacturer: Obuchow
Designation: 7.6cm Mountain gun M 1904
Calibre (mm): 76.2
Length of barrel (x calibre): 13.3*
Depression/Elevation (^{o}): -10 +25
Traverse (^{o}): —
Weight in firing position (kg): 327
Weight of projectile (kg): 6.5
Muzzle velocity (m/sec): 295*
Maximum range (km): 4.16*
Gun carriage: Box trail
Gun shield: Yes
Mode of traction: Horse drawn, pack animals
Introduced into: Russia

Remarks:

Dismantles into five parts. Replacement by the 75mm Mountain Gun M 09 (Schneider) was planned before World War I, but only partly carried out.

 *Using the same gun carriage with an M 1909 barrel of greater length— muzzle velocity 380 m/sec, range 6.4km.

Country of origin: Soviet Union
Manufacturer: state
Designation: 76mm Mountain Gun M-1938
Calibre (mm): 76.2
Length of barrel (x calibre): 23
Depression/Elevation (°): -8 +70
Traverse (°): 10
Weight in firing position (kg): 785
Weight of projectile (kg): 6.23
Muzzle velocity (m/sec): 495
Maximum range (km): 10.1
Gun carriage: Box trail
Gun shield: Yes
Mode of traction: Motor drawn, by pack animals
Introduced into: Soviet Union

Remarks:

Dismantles into ten parts. During World War II this gun replaced the
75mm Mountain Gun M 09 (Schneider), and the 105mm Mountain
Howitzer M 09 (Schneider). Designed by the firm of Škoda, from where
the plans of the 75mm Mountain Gun M 36 (C 5) were bought in 1936.
Prototypes with a calibre of 76.2mm were still in production at Škoda.

Fieldgun 75mm   **Spain**

Country of origin: Spain (?)
Manufacturer: —
Designation: Gun 75/22
Calibre (mm): 75
Length of barrel (x calibre): 22
Depression/Elevation (°): -0 +40
Traverse (°): 7
Weight in firing position (kg): 764
Weight of projectile (kg): —
Muzzle velocity (m/sec): 450
Maximum range (km): 10.5
Gun carriage: Box trail
Gun shield: Yes
Mode of traction: Horse drawn
Introduced into: Spain

Remarks:

Country of origin: Spain (?)
Manufacturer: —
Designation: Howitzer 105/26
Calibre (mm): 105
Length of barrel (x calibre): 26
Depression/Elevation (^{o}): -5 +45
Traverse (^{o}): 50
Weight in firing position (kg): 1,950
Weight of projectile (kg): —
Muzzle velocity (m/sec): 443
Maximum range (km): 9.4
Gun carriage: Split trail
Gun shield: Yes
Mode of traction: Motor drawn
Introduced into: Spain

Remarks:

Country of origin: Sweden/Germany
Manufacturer: Bofors/Krupp
Designation: 7.5cm Fieldgun M 02/33
Calibre (mm): 75
Length of barrel (x calibre): 30
Depression/Elevation (^{o}): -5 +42
Traverse (^{o}): 50
Weight in firing position (kg): 1,350
Weight of projectile (kg): 6.5
Muzzle velocity (m/sec): 540
Maximum range (km): 11.0
Gun carriage: Split trail
Gun shield: Yes
Mode of traction: Horse drawn
Introduced into: Sweden

Remarks:

Barrel of the old 75mm Fieldgun M 02 (Krupp) in a new split trail gun
carriage made by Bofors.

Country of origin: Sweden
Manufacturer: Bofors
Designation: 7.5cm Fieldgun M 40
Calibre (mm): 75
Length of barrel (x calibre): 40
Depression/Elevation (^{o}): -5 +45
Traverse (^{o}): 50
Weight in firing position (kg): 1,500
Weight of projectile (kg): 6.3
Muzzle velocity (m/sec): 700
Maximum range (km): 14.0
Gun carriage: Split trail
Gun shield: Yes
Mode of traction: Horse and motor drawn
Introduced into: Sweden

Remarks:
Standard gun carriage with the 105mm Field Howitzer M 40 (Bofors). The fieldgun was replaced by the field howitzer.

Field Howitzer 105mm M 10 Sweden

Country of origin: Sweden
Manufacturer: Bofors
Designation: 10.5cm Light Field Howitzer M 10
Calibre (mm): 105
Length of barrel (x calibre): 16
Depression/Elevation (^{O}): -5 +43
Traverse (^{O}): 4
Weight in firing position (kg): 1,225
Weight of projectile (kg): 14.0
Muzzle velocity (m/sec): 304
Maximum range (km): 6.2
Gun carriage: Box trail
Gun shield: Yes
Mode of traction: Horse drawn
Introduced into: Sweden

Remarks:
Replaced by the 105mm Field Howitzer M 40 (Bofors).

Country of origin: Sweden
Manufacturer: Bofors
Designation: 10.5cm Gun
Calibre (mm): 105
Length of barrel (x calibre): 22
Depression/Elevation (°): -5 +45
Traverse (°): 8.5 (360 *
Weight in firing position (kg): 1,650
Weight of projectile (kg): 14.0
Muzzle velocity (m/sec): 475
Maximum range (km): 10.5
Gun carriage: Box trail
Gun shield: Yes
Mode of traction: Horse and motor drawn
Introduced into: Netherlands (Colonies)

Remarks:

Introduced between the two world wars into the artillery units of the Dutch Colonies. Not used in The Netherlands.

*With wheel rim (as illustrated).

Field Howitzer 105mm M 40

Sweden

Country of origin: Sweden
Manufacturer: Bofors
Designation: 10.5cm Howitzer M 40
Calibre (mm): 105
Length of barrel (x calibre): 22
Depression/Elevation (°): -5 +45
Traverse (°): 50
Weight in firing position (kg): 1,840
Weight of projectile (kg): 15.4
Muzzle velocity (m/sec): 460
Maximum range (km): 10.0
Gun carriage: Split trail
Gun shield: Yes
Mode of traction: Horse or motor drawn
Introduced into: Sweden

Remarks:

Replaced the 105mm Field Howitzer M 10 (Bofors).

Country of origin: Sweden
Manufacturer: Bofors
Designation: 10.5cm Howitzer 4140
Calibre (mm): 105
Length of barrel (x calibre): 28
Depression/Elevation (°): -5 +65
Traverse (°): 360
Weight in firing position (kg): 2,600
Weight of projectile (kg): 15.2
Muzzle velocity (m/sec): 610
Maximum range (km): 14.6
Gun carriage: Quadruple spar
Gun shield: Yes
Mode of traction: Motor drawn
Introduced into: Sweden

Remarks:
Replaced the older models in the Infantry Brigades. Mainly drawn by
tractors. The first fieldgun to be series-produced with a traverse of 360°.

Country of origin: Sweden
Manufacturer: Bofors
Designation: 7.5cm Mountain Gun M 30
Calibre (mm): 75
Length of barrel (x calibre): 20
Depression/Elevation (^{o}): -10 +50
Traverse (^{o}): 6
Weight in firing position (kg): 800
Weight of projectile (kg): 6.5
Muzzle velocity (m/sec): 405
Maximum range (km): 9.2
Gun carriage: Box trail
Gun shield: Yes
Mode of traction: Horse drawn, pack animals
Introduced into: Turkey

Remarks:
Dismantles into eight parts.

Country of origin: Sweden
Manufacturer: Bofors
Designation: 7.5cm Mountain Gun M 36
Calibre (mm): 75
Length of barrel (x calibre): 20
Depression/Elevation (°): -10 +50
Traverse (°): 5
Weight in firing position (kg): 775
Weight of projectile (kg): 6.5
Muzzle velocity (m/sec): 400
Maximum range (km): 8.2
Gun carriage: Box trail
Gun shield: Yes
Mode of traction: Horse drawn, pack animals
Introduced into: Argentina, Bulgaria

Remarks:
Dismantles into eight parts. Also known as the Bofors/Krupp.

Fieldgun 75mm M 03/22   Switzerland

Country of origin: Switzerland/Germany
Manufacturer: Sulzer/Krupp
Designation: 7.5cm Fieldgun
Calibre (mm): 75
Length of barrel (x calibre): 30
Depression/Elevation (^{O}): -8 +25.5*
Traverse (^{O}): 8
Weight in firing position (kg): 1,096
Weight of projectile (kg): 6.35
Muzzle velocity (m/sec): 485
Maximum range (km): 10.0*
Gun carriage: Box trail
Gun shield: Yes
Mode of traction: Horse drawn
Introduced into: Switzerland

Remarks:

Rebuilt in Switzerland. Dismantles into nine parts for mountain transportation. Rebuild of the 75mm Fieldgun M 03 (Krupp) in 1923/24. Main weapon of the Swiss Field Artillery until World War II. 75mm Fieldgun M 03/40 for the Light Brigades with a new split trail gun carriage. Motor drawn.

* Jacked up elevation can be increased to +41°, increasing range to 11 km.

Country of origin: Switzerland
Manufacturer: K. and W. Thun
Designation: 10.5cm Howitzer 46
Calibre (mm): 105
Length of barrel (x calibre): 22
Depression/Elevation (^{O}): -0 +65 *
Traverse (^{O}): —
Weight in firing position (kg): 1,850
Weight of projectile (kg): 15.15
Muzzle velocity (m/sec): 490
Maximum range (km): 10.0
Gun carriage: Split trail
Gun shield: Yes
Mode of traction: Motor drawn
Introduced into: Switzerland

Remarks:

Main weapon of the Howitzer units of the divisional artillery. Replaced the 120mm Field Howitzer M 12 (Krupp). Produced from 1943 until 1953.

* Partly? only +45°.

Country of origin: Switzerland/Sweden
Manufacturer: state/Bofors
Designation: 7.5cm Mountain Gun 33
Calibre (mm): 75
Length of barrel (x calibre): 22
Depression/Elevation (^{o}): -10 +50
Traverse (^{o}): 6
Weight in firing position (kg): 790
Weight of projectile (kg): 6.5
Muzzle velocity (m/sec): 500
Maximum range (km): 10.5
Gun carriage: Box trail
Gun shield: Yes
Mode of traction: Horse or motor drawn, pack animals
Introduced into: Switzerland

Remarks:

Dismantles into eight pieces, or two sections tandem drawn. Built in
Switzerland under licence. Replaced the obsolete 75mm Mountain Gun
M 06 (Krupp). In use until after World war II. Designed for fieldgun
ammunition.

Modified version of the 75mm M 33 Mountain Gun with pneumatic tyres after World War II.

Country of origin: USA
Manufacturer: Ordnance Department
Designation: 7.62cm (3'') Fieldgun M 02
Calibre (mm): 76.2
Length of barrel (x calibre): 29.2
Depression/Elevation (°): -5 + 15
Traverse (°): 8
Weight in firing position (kg): 970
Weight of projectile (kg): 6.8
Muzzle velocity (m/sec): 518
Maximum range (km): 6.86/8.65 *
Gun carriage: Box trail
Gun shield: Yes
Mode of traction: Horse drawn
Introduced into: USA

Remarks:

Introduced in 1905 after extensive trials. Those produced, however, were
only used by the peacetime army. Replaced by the French 75mm Fieldgun
M 97 (Schneider), and the American 75mm Fieldguns M 16 (Ordnance
Department) and M 17 (Bethlehem). 50 guns were manufactured by
Ehrhardt. Similar to the 96.5mm Fieldgun M 07 (Ordnance Department),
barrel length 27.1 calibre, projectile weight 13.6 kg, muzzle velocity 518
m/sec, range 7.3 km.

*Normal/with entrenched trail spade.

Country of origin: USA
Manufacturer: Ordnance Department
Designation: 75mm Fieldgun M 16
Calibre (mm): 75
Length of barrel (x calibre): 30.8
Depression/Elevation (°): -7 +53
Traverse (°): 45
Weight in firing position (kg): 1,380
Weight of projectile (kg): 6.12
Muzzle velocity (m/sec): 600/630 *
Maximum range (km): 8.8/11.4 *
Gun carriage: Split trail
Gun shield: Yes
Mode of traction: Horse or motor drawn
Introduced into: USA

Remarks:

Since 1912 attempts have been made to replace the 76.2mm Fieldgun M 02 (Ordnance Department). At the end of World War I there were only 810 of these guns in existence, compared with 3,854 of the French 75mm Fieldgun M 97 (Schneider). The original calibre of 76.2mm was changed to 75mm to enable the gun to fire the ammunition of the French 75mm Fieldgun.

* Old/new ammunition from 1918.

Country of origin: USA
Manufacturer: Bethlehem
Designation: 75mm Fieldgun M 17
Calibre (mm): 75
Length of barrel (x calibre): 30
Depression/Elevation (°): -5 +16
Traverse (°): 8
Weight in firing position (kg): 1,310
Weight of projectile (kg): 6.8
Muzzle velocity (m/sec): 531/565 *
Maximum range (km): 8.0/12.3 *
Gun carriage: Pole trail
Gun shield: Yes
Mode of traction: Horse drawn
Introduced into: USA

Remarks:

British design. At the end of World War I only 909 of this type were in existence, as against, 3,854 of the French 75mm Fieldgun M 97 (Schneider). The gun was ordered by Bethlehem from Great Britain. However the contract was ceded to the USA.

* With old/new ammuniton.

Country of origin: USA/France
Manufacturer: Ordnance Department/Schneider
Designation: 7.5cm Fieldgun M 97A4
Calibre (mm): 75
Length of barrel (x calibre): 36
Depression/Elevation (^{o}): -10 +19
Traverse (^{o}): 6
Weight in firing position (kg): 1,364
Weight of projectile (kg): 6.2
Muzzle velocity (m/sec): 575
Maximum range (km): 11.2
Gun carriage: Box trail
Gun shield: Yes
Mode of traction: Motor drawn
Introduced into: USA

Remarks:

The main weapon of the American Artillery until the beginning of World
War II. An American gun carriage was fitted to the barrel of the French
75mm Fieldgun M 97 (Schneider).

Fieldgun 75mm M 2A2

Country of origin: USA
Manufacturer: Ordnance Department
Designation: 7.5cm Fieldgun M 2A2
Calibre (mm): 75
Length of barrel (x calibre): 40
Depression/Elevation (°): -10 +46
Traverse (°): 85
Weight in firing position (kg): 1,563
Weight of projectile (kg): 6.8
Muzzle velocity (m/sec): 663
Maximum range (km): 13.7
Gun carriage: Split trail
Gun shield: No
Mode of traction: Motor drawn
Introduced into: USA

Remarks:

Intended to be the standard equipment of the divisional artillery but replaced by the field howitzer.

Country of origin: USA
Manufacturer: Ordnance Department
Designation: 7.5cm Light Field Howitzer M-3A1
Calibre (mm): 75
Length of barrel (x calibre): 19.1
Depression/Elevation (°): -9 +49
Traverse (°): 55
Weight in firing position (kg): 853
Weight of projectile (kg): 6.67
Muzzle velocity (m/sec): 381
Maximum range (km): 8.79
Gun carriage: Split trail
Gun shield: No
Mode of traction: Motor drawn
Introduced into: USA

Remarks:
Developed before World War II. Only produced in small numbers. The gun
carriage was later used for the 105mm Field Howitzer M-3A1 (Ordnance
Department).

Country of origin: USA
Manufacturer: Ordnance Department
Designation: 3.8″ Howitzer M 17
Calibre (mm): 96.5
Length of barrel (x calibre): 22.6
Depression/Elevation (°): -5 +48
Traverse (°): 45
Weight in firing position (kg): —
Weight of projectile (kg): 13.6
Muzzle velocity (m/sec): 366
Maximum range (km): 7.8
Gun carriage: Split trail
Gun shield: —
Mode of traction: —
Introduced into: USA

Remarks:

Tested in 1913. Only eight batteries were ordered.

Country of origin: USA
Manufacturer: Ordnance Department
Designation: 105mm Howitzer M-2A1
Calibre (mm): 105
Length of barrel (x calibre): 22
Depression/Elevation (°): -5 +66
Traverse (°): 46
Weight in firing position (kg): 2,030/2,259 *
Weight of projectile (kg): 14.97
Muzzle velocity (m/sec): 472
Maximum range (km): 11.2
Gun carriage: Split trail
Gun shield: Yes
Mode of traction: Motor drawn
Introduced into: USA, numerous other states

Remarks:
After modifications became known as M-101 or M-101A1. 8,536 guns of this design were delivered between July 1940 and August 1945. The standard equipment of the American divisional artillery from the beginning of World War II.

* In gun carriage M-2A1/M-2A2.

Country of origin: USA
Manufacturer: Ordnance Department
Designation: 105mm Howitzer M-3A1
Calibre (mm): 105
Length of barrel (x calibre): 18
Depression/Elevation (°): -9 +49
Traverse (°): 55
Weight in firing position (kg): 1,134
Weight of projectile (kg): 14.97
Muzzle velocity (m/sec): —
Maximum range (km): 6.4
Gun carriage: Split trail
Gun shield: No
Mode of traction: Motor drawn
Introduced into: USA

Remarks:

Introduced in World War II. 2,580 guns were delivered up to 1945.
Intended for airborne units and for jungle warfare. The shortened barrel of
the 105mm Field Howitzer M-2A1 (Ordnance Department) was placed in
the gun carriage of the 75mm Field Howitzer M-3A1 (Ordnance
Department).

Country of origin: USA
Manufacturer: Ordnance Department
Designation: 105mm Light field Howitzer M-102
Calibre (mm): 105
Length of barrel (x calibre): —
Depression/Elevation (^{0}): -5 +76
Traverse (^{0}): 360
Weight in firing position (kg): 1,362
Weight of projectile (kg): 12.7*
Muzzle velocity (m/sec): —
Maximum range (km): 15.1
Gun carriage: Box trail
Gun shield: No
Mode of traction: Motor drawn
Introduced into: USA

Remarks:
Introduced into the airborne units about 1965.

*With the ammunition of the 105mm Field Howitzer M-2A1, 14.97 kg, maximum range 11.0 km.

Country of origin: USA
Manufacturer: Ordnance Department
Designation: 4.7'' Howitzer M 1908 M1
Calibre (mm): 119.4
Length of barrel (x calibre): —
Depression/Elevation (°): -5 +40
Traverse (°): —
Weight in firing position (kg): 2,177*
Weight of projectile (kg): 27.2
Muzzle velocity (m/sec): 274
Maximum range (km): 6.26
Gun carriage: Box trail
Gun shield: Yes
Mode of traction: Horse drawn
Introduced into: USA

Remarks:
Produced only in small numbers until about 1915.

*In travelling position.

Country of origin: USA
Manufacturer: —
Designation: 3″ Mountain Howitzer M 11
Calibre (mm): 76.2
Length of barrel (x calibre): —
Depression/Elevation (^{o}): -5 +40
Traverse (^{o}): 6
Weight in firing position (kg): 505
Weight of projectile (kg): 6.8
Muzzle velocity (m/sec): 274
Maximum range (km): 5.2
Gun carriage: —
Gun shield: Yes
Mode of traction: Horse drawn, pack animals
Introduced into: USA

Remarks:

Dismantles into five parts.

Country of origin: USA
Manufacturer: Ordnance Department
Designation: 75mm Mountain Howitzer M 1
Calibre (mm): 75
Length of barrel (x calibre): 19
Depression/Elevation (^{o}): -5 +45
Traverse (^{o}): 5
Weight in firing position (kg): 576
Weight of projectile (kg): 6.8
Muzzle velocity (m/sec): 381
Maximum range (km): 8.4
Gun carriage: Box trail
Gun shield: No
Mode of traction: Horse drawn, pack animals
Introduced into: USA

Remarks:

Dismantles into six parts. After various prototypes at the beginning of the twenties this mountain howitzer was finally introduced in 1928.

Mountain Howitzer 75mm M-1A1

Country of origin: USA
Manufacturer: Ordnance Department
Designation: 75mm Mountain Howitzer M-1A1
Calibre (mm): 75
Length of barrel (x calibre): 19
Depression/Elevation (°): -5 +45
Traverse (°): 6
Weight in firing position (kg): 653
Weight of projectile (kg): 6.67
Muzzle velocity (m/sec): 381
Maximum range (km): 8.79
Gun carriage: Box trail
Gun shield: No
Mode of traction: Motor drawn, pack animals
Introduced into: USA, Great Britain

Remarks:

Dismantles into seven parts. After modification M-116. Development of the old 75mm Mountain Howitzer M 1 (Ordnance Department). Used by the airborne and Marine troops. 4,939 guns were delivered during world War II.

Field Howitzer 105mm XM-164. Experimental gun made up from Field Howitzer M-101 A1 and M-102 utilizing aluminium alloy. Development by the Marine Corps.

Field Howitzer 105mm, XM-204. Experimental gun with soft recoil.

Country of origin: Yugoslavia
Manufacturer: state
Designation: 105mm Field Howitzer M-56
Calibre (mm): 105
Length of barrel (x calibre): —
Depression/Elevation (°): -12 +68
Traverse (°): 52
Weight in firing position (kg): 2,100
Weight of projectile (kg): 14.97
Muzzle velocity (m/sec): 570
Maximum range (km): 13.0
Gun carriage: Split trail
Gun shield: Yes
Mode of traction: Motor drawn
Introduced into: Yugoslavia

Remarks:

Modified version of the American 105mm Field Howitzer M-2A1
(Ordnance Department).

Country of origin: Yugoslavia
Manufacturer: state
Designation: 76mm Mountain Gun M-48 B-1
Calibre (mm): 76
Length of barrel (x calibre): 15.5
Depression/Elevation (°): -15 +45
Traverse (°): 50
Weight in firing position (kg): 705
Weight of projectile (kg): 6.2
Muzzle velocity (m/sec): 398
Maximum range (km): 8.6
Gun carriage: Split trail
Gun shield: Yes
Mode of traction: Horse and motor drawn, by pack animals
Introduced into: Yugoslavia

Remarks:

Also used as an Infantry gun. The M-48B-1A1 and M 48B-1A2 dismantle into three parts.

INDEX

List of the guns included in the Tables of Type
(Arranged under type of gun, calibre, manufacturer and mark)

Type of Gun	Calibre (mm)	Manufacturer	Mark	Page
Field Howitzers	120	Rheinmetall	M 09	145
		Schneider	M 97	97
		Schneider	M 07	99
		Schneider	M 11	99
		Schneider	M 15	100
	122	Krupp	M 09	142
		Putilow/Obuchow	M 04	206
		Schneider	M 10	98
		state/Schneider	M 10/30	207
		state/Krupp	M 09/37	208
		state/Soviet Union	M 38	209
	127	./Great Britain	M 97	167
Fieldguns	75	Ansaldo	M 37	174
		Arisaka/Osaka Arsenal	M 05	180
		Osaka Arsenal/Krupp	M 08	181
		Osaka Arsenal	M 05 improved	182
		Osaka Arsenal	M 30	183
		Osaka Arsenal	M 35	184
		Bethlehem	M 17	229
		Bofors/Krupp	M 02/33	215
		Bofors	M 40	216
		Cockerill/Rheinmetall	GP I	61
		Cockerill/Rheinmetall	GP II	62
		Cockerill/Rheinmetall	GP III	63
		Deport	M 11	80
		Ehrhardt	M 01	115
		FRC/Krupp	Quickfiring	60
		Krupp	M 02	116
		Krupp	M 02	117
		Krupp	M 03	118
		Krupp	M 03	119
		Krupp	M 03	120
		Krupp	M 03	121
		Krupp	M 05	123
		Krupp	M 05	124
		Krupp	M 06	125
		Krupp	M 11	126
		Krupp	M 16 New Type	128
		Krupp	M 18	129
		Krupp	M 38	130
		Ordnance Department	M 16	228
		Ordnance Department	M-2A2	231
		Ordnance Department/ Schneider	M 97A4	230
		Rheinmetall	7 M 85	132
		St Chamond		81
		Schneider	M 97	81
		Schneider	M 03	84
		Schneider	M 05	85
		Schneider	M 06	86
		Schneider	M 06 (PDM)	87

Type of Gun	Calibre (mm)	Manufacturer	Mark	Page
Fieldguns	75	Schneider	M 12	88
		Schneider	M 14	88
		Schneider	M 22	89
		Schneider	M 97/33	83
		Siderius/Krupp	M 02/04 vd OM	188
		Siderius/Krupp	M 02/04 vd	189
		Siderius/Krupp	M 02/04 vd NM	190
		Škoda	M 11	45
		Škoda	M 28 (EFI	64
		Škoda	M 35 (E3)	66
		./Spain		213
		state/Putilow	M 02/26	195
		Sulzer/Krupp	M 03/22	223
		Krupp	M 03/40	223
	76.2	Ehrhardt	M 00	114
		Ordnance Department	M 02	227
		Putilow	M 00	193
		Putilow	M 02	194
		state/Soviet Union	M 02/30	196
		state/Soviet Union	M 33	197
		state/Soviet Union	M 36	198
		state/Soviet Union	M 39 (USW)	199
		state/Soviet Union	M 41	200
		state/Soviet Union	M 39/42	201
		state/Soviet Union	M 42 (S1S-3)	202
		Vickers	M 04	158
	76.5	Böhler	M 18	42
		Škoda	M 17	46
		Škoda	M 30 (NPK)	65
		Škoda	M 39 (E7)	67
		state/Škoda	M 05	43
		state/Škoda	M 05/08	44
	77	Krupp/Ehrhardt	M 96 New Type	122
		Krupp	M 96/16	127
		Rheinmetall	M 16	131
	83.8	Vickers	M 03	159
		Vickers	Mk II	159
		Vickers	Mk IV	160
		Vickers	Mk V	161
	85	state/Soviet Union	M 43	203
		state/Soviet Union	M 45 (D-44)	204
		state/Soviet Union	M 45 (D-48)	205
		state Czechoslovakia	M 52	68
	90	de Bange	M 77	90
Mountain	96.5	Ordnance Department	M 07	227
Howitzers	75	Ordnance Department	M 1	239
		Ordnance Department	M-1A1 (M-116)	240
		Vickers	M 01	169
	76.2	./USA	M 11	238
	90	Škoda	M 28 (DC)	75
	94	Vickers	M 18	170

Type of Gun	Calibre (mm)	Manufacturer	Mark	Page
Mountain Howitzers	100	Škoda	M 16	58
		Škoda	M 16/19	76
	104	Škoda	M 08	56
		Škoda	M 10	57
	105	Böhler	M 40	156
		Krupp		157
	105	OTO Melara	M 56	179
		Royal Armaments Research Establishment		171
		Schneider	M 09	111
		Schneider	M 19	112
		Schneider	M 28	
		Škoda	M 16(T)	58
		Škoda	M 39 (D8)	77
		Škoda	M 39 (D9)	77
	150	Škoda	M 18	59
Mountain Guns	37	Krupp		146
	65	Turin Arsenal	M 13	176
		Schneider-Ducrest	M 06	101
	70	Turin Arsenal	M 02	177
		Schneider	M 07	102
		Schneider	M 07	103
		Schneider	M 08 (MD)	104
		Vickers	M 11	168
	72.5	Artillerie-Zeugs-Fabr	M 08	52
		Škoda	M 09	53
		state/Austria-Hungary	M 99	51
	75	Ansaldo	M 34	178
		Osaka Arsenal	M 08 mod.	186
		Osaka Arsenal	M 34	187
		Bofors	M 30	221
		Bofors	M 36	222
		Ehrhardt	M 10	147
		Kongsberg	M 27	192
		Krupp	M 04	148
		Krupp	M 05	148
		Krupp	M 06	149
		Krupp	M 08	150
		Krupp	M 13	151
		MÁVAG/Škoda	M 15/31	173
		MÁVAG/Škoda	M 15/35	173
		Rheinmetall	M 08	152
		Rehinmetall	M 14	153
		Rheinmetall	M 36	155
		Schneider	M 07	105
		Schneider	(MD)	106
		Schneider	M 19	109
		Schneider	M 28	110
		Schneider-Danglis	M 06/09	107
		Škoda	M 13	54
		Škoda	M 15	55

Type of Gun	Calibre (mm)	Manufacturer	Mark	Page
Mountain Guns	75	Škoda	M 28 (CD)	73
		Škoda	M 39 (C6)	74
		state/Bofors	M 33	225
		state/Bofors	M 38	226
	76.2	Obuchow	M 04	211
		Schneider	M 09	108
		state/Yugoslavia	M 48	243
		state/Soviet Union	M 38	212
	77	Rheinmetall	M 15	154
Gun Howitzers	85	Schneider	M 27	91
	87.6	state/Vickers	Mk I	162
		state/Great Britain	Mk II	163
	122	state/Soviet Union	M 63 (D-30)	210